BUSINESS ISSUES, COMPETITION AND ENTREPRENEURSHIP

RESTORING AMERICAN COMPETITIVENESS AND INNOVATION

BUSINESS ISSUES, COMPETITION AND ENTREPRENEURSHIP

Additional books in this series can be found on Nova's website under the Series tab.

Additional E-books in this series can be found on Nova's website under the E-book tab.

ECONOMIC ISSUES, PROBLEMS AND PERSPECTIVES

Additional books in this series can be found on Nova's website under the Series tab.

Additional E-books in this series can be found on Nova's website under the E-book tab.

BUSINESS ISSUES, COMPETITION AND ENTREPRENEURSHIP

RESTORING AMERICAN COMPETITIVENESS AND INNOVATION

AMBER J. THOMAS
AND
NELSON R. RAMIREZ
EDITORS

Nova Science Publishers, Inc.
New York

For permission to use material from this book please contact us:
Telephone 631-231-7269; Fax 631-231-8175
Web Site: http://www.novapublishers.com

Library of Congress Cataloging-in-Publication Data

ISBN: 978-1-62081-509-0

Published by Nova Science Publishers, Inc. † New York

CONTENTS

PREFACE

The U.S. economy reigned supreme in the 20th century, becoming the largest most productive, and most competitive in the world; amazing new technologies were invented and commercialized; the workforce became the most educated in the world; and incomes soared while a large middle class emerged and thrived. As the 21st century approached, however, alarms began to sound about the U.S. economy's ability to remain in this preeminent position. Incomes stagnated and job growth slowed. Other countries became better educated and our manufacturing sector lost ground to foreign competitors. Observers have expressed concern that the scientific and technological building blocks critical to our economic leadership have been eroding at a time when many other nations are actively laying strong foundations in these same areas. This book provides an overview of the restoration of American innovation which is the key driver of competitiveness, wage and job growth, and long-term economic growth.

In: Restoring American Competitiveness ... ISBN: 978-1-62081-509-0
Editors: A.J.Thomas and N.R.Ramirez

Chapter 1

THE COMPETITIVENESS AND INNOVATIVE CAPACITY OF THE UNITED STATES*

United States Department of Commerce

FOREWORD

On January 4, 2011, President Barack Obama signed into law the America COMPETES Reauthorization Act of 2010 (COMPETES). Section 604 of COMPETES mandates that the Secretary of Commerce complete a study that addresses the economic competitiveness and innovative capacity of the United States (see Supplemental Materials). Congress directed that this report address a diverse array of topics and policy options, including: tax policy; the general business climate in the U.S.; regional issues such as the role of state and local governments in higher education; barriers to setting up new firms; trade policy, including export promotion; the effectiveness of Federal research and development policy; intellectual property regimes in the U.S. and abroad; the health of the manufacturing sector; and science and technology education.

In conducting this study, COMPETES specified that the Secretary of Commerce establish a process for obtaining comments. One part of that process was to establish a 15 member Innovation Advisory Board (IAB) "for

---

* This is an edited, reformatted and augmented version of United States Department of Commerce Publication, in consultation with the National Economic Council, dated January 2012.

purposes of obtaining advice with respect to the conduct of the study." The Department of Commerce announced the members of the IAB (listed in the Supplementary Materials section of this report) on May 4, 2011, and the inaugural meeting of the IAB was on June 6, 2011, in Alexandria, Virginia. A second meeting of the IAB was held September 23, 2011, in Boulder, Colorado. IAB members provided input into the process throughout the summer. Additionally, some IAB members generously hosted COMPETES-related events in Washington, D.C.; Youngstown, Ohio; Morgantown, West Virginia; Philadelphia, PA; and New York, NY. These events brought together community and business leaders, and experts in a wide variety of areas, to share their ideas on competitiveness. Department of Commerce and Administration staff attended all of these meetings.

Additionally, we received input from a number of other groups at various events. These included an all day event with a group of prominent academic economists in Cambridge, Massachusetts, and a conference at the Silicon Flatirons' Center for Law, Technology, and Entrepreneurship at the University of Colorado. Other groups, as well as the general public, provided additional input.

We are very grateful for the generosity of all contributors, but special thanks go to the Innovation Advisory Board members—they passionately care about the future of this country and have been willing to give their valuable time and expertise to enrich this process.

Sincerely,

John E. Bryson
Secretary of Commerce

EXECUTIVE SUMMARY

The U.S. economy reigned supreme in the 20th century, becoming the largest, most productive, and most competitive in the world; amazing new technologies were invented and commercialized; the workforce became the most educated in the world; and incomes soared while a large middle class emerged and thrived. As the 21st century approached, however, alarms began to sound about the U.S. economy's ability to remain in this preeminent position. Incomes stagnated and job growth slowed. Other countries became better educated and our manufacturing sector lost ground to foreign competitors. Observers have expressed concern that the scientific and technological building blocks critical to our economic leadership have been eroding at a time when many other nations are actively laying strong foundations in these same areas. In short, some elements of the U.S. economy are losing their competitive edge which may mean that future generations of Americans will not enjoy a higher standard of living than is enjoyed in the United States today.

Innovation is the key driver of competitiveness, wage and job growth, and longterm economic growth. Therefore, one way to approach the question of how to improve the competitiveness of the United States is to look to the past and examine the factors that helped unleash the tremendous innovative potential of the private sector. Among these factors, three pillars have been key: Federal support for basic research, education, and infrastructure. Federally supported research laid the groundwork for the integrated circuit and the subsequent computer industry; the Internet; and advances in chemicals, agriculture, and medical science. Millions of workers can trace their industries and companies back to technological breakthroughs funded by the government. The U.S. educational system in the 20th century produced increasing numbers of high school and college graduates, more so than anywhere else in the world. These highly skilled workers, in turn, boosted innovation. The transformation of infrastructure in the 20th century was nothing short of amazing: the country became electrified, clean water became widely available, air transport became ubiquitous, and the interstate highway system was planned and constructed. All of these developments helped businesses compete by opening up markets and keeping costs low.

Common to all three pillars—research, education, and infrastructure—is that they are areas where government has made, and should continue to make, significant investments. For a variety of reasons, the private sector underinvests in these areas so the government needs to step in to bring investment up to the socially optimal levels. An additional common thread between these three pillars is that the benefits of these investments took years to be fully realized. For instance, we are still benefiting today from investments made in the 19th century, such as the Morrill Act of 1862, which laid the foundation for the land grant university system in all states. In the 20th century, World War II-era research became the basis of the transistor; and in the 1960s, all of the benefits from investing in science made the United States the leader of the space race as well as the information technology industry. This long-term outlook should not be forgotten.

The need for the Federal government to play an important role in the first pillar—research, particularly basic research–derives from the fact that there is a divergence between the private and social returns of research activities which leads to less innovative activity in the private sector than is what is best for our country. However, government support of basic research can remedy this problem. The benefits from Federal research and development (R&D) support are not just theoretical: as mentioned above, the Federal government has played a crucial role in the development of many key innovations of the mid- to late-20th century.

Federal funding for basic research has been increasing, but at a slower pace than economic growth. To improve the trajectory of American innovation, thoughtful, decisive, and targeted actions are needed, some of which already have been proposed. These actions include sustaining the levels of funding for basic research by the Federal government, extending a tax credit for private-sector R&D to give companies appropriate and well-designed incentives to boost innovation above the baseline level that would have been reached absent these incentives, and improving the methods by which basic research is transferred from the lab into commercial products.

Education, the second pillar, is also critical to foster innovation and to increase living standards. The advances in education in the 20th century helped propel the economic rise of the United States as it became the richest nation on the planet. However, by many measures, the U.S. education system has slipped. By some accounts, the United States' system of higher education remains the best in the world and educates our country's and our competitors' future scientists and engineers, factors such as poor preparation in math and science and the high cost of college tuition and expenses are restricting the flow of American science, technology, engineering and mathematics (STEM) graduates from our universities.

Ongoing and new Administration initiatives are addressing these challenges by making college more affordable, spurring classroom innovation at all levels, expanding the size and quality of the STEM teacher ranks, and encouraging and facilitating students' and workers' continued STEM education.

In the past, the United States led the way in several key areas of infrastructure development, the third pillar of innovative capacity, starting with the railroad system of the 1800's. In today's economy, the nature of infrastructure needed to compete is changing and the United States is lagging behind in certain key aspects of a 21st century infrastructure (such as broadband Internet access) and facing capacity constraints for other aspects (wireless communications) given the high demand for these services. Ensuring that the United States has the infrastructure it needs to be competitive in the 21st century will require both additional support by the government and an appropriate policy framework to enable the private sector to build on the government's support.

A crucial component of the United States' future competitive strength is a flourishing manufacturing sector. Manufacturing creates high-paying jobs, provides the bulk of U.S. exports, and spurs innovation. While manufacturing continues to play a vital role in the U.S. economy and provides jobs for millions of Americans, it also has faced significant challenges, especially over the last decade. Manufacturing's share of GDP and the number of workers in manufacturing has fallen, while the trade balance in manufactured goods has worsened. In the manufacturing sector, the Federal government has historically played an important role in providing a level playing field and must do so with renewed vigor to ensure that manufacturing continues to thrive in the United States. The current and future health of the manufacturing sector is strongly linked to the investments we make in research, education, and infrastructure.

Increasing the competitiveness and the capacity to innovate goes beyond improving research, education, infrastructure and manufacturing. Many other policies that ensure the private sector has the best possible environment in which to innovate contribute to competitiveness, including incentives to form regional clusters, promotion of exports and access to foreign markets, the level and structure of corporate taxes, and an effective intellectual property regime (domestically and abroad). In each of these areas, the Federal government has an important role to play.

The challenges are great, but the United States has a strong base on which to build and to rise to these challenges. There are clear actions that can help this nation regain its innovative and competitive footing. To succeed, we must have the will to implement and to sustain the policies that will prepare the United States to continue to be an economic leader in the 21st century.

RISING TO THE CHALLENGE

The U.S. economy was the world leader in the 20th century; moving into the 21st century, however, various parties have raised alarms about whether this nation's economy can continue to be competitive. The U.S. economy remains the largest in the world, possessing a highly skilled work force, world class companies, and, according to some, the world's best higher education system. Despite these positive attributes, U.S. citizens have been hit by stagnating job growth and falling incomes, while businesses have faced increasing global competition. In short, the concern is that future generations of Americans will not enjoy a higher standard of living than is enjoyed today. With the right policies and commitment, the United States can compete globally and provide its citizens with better lives.

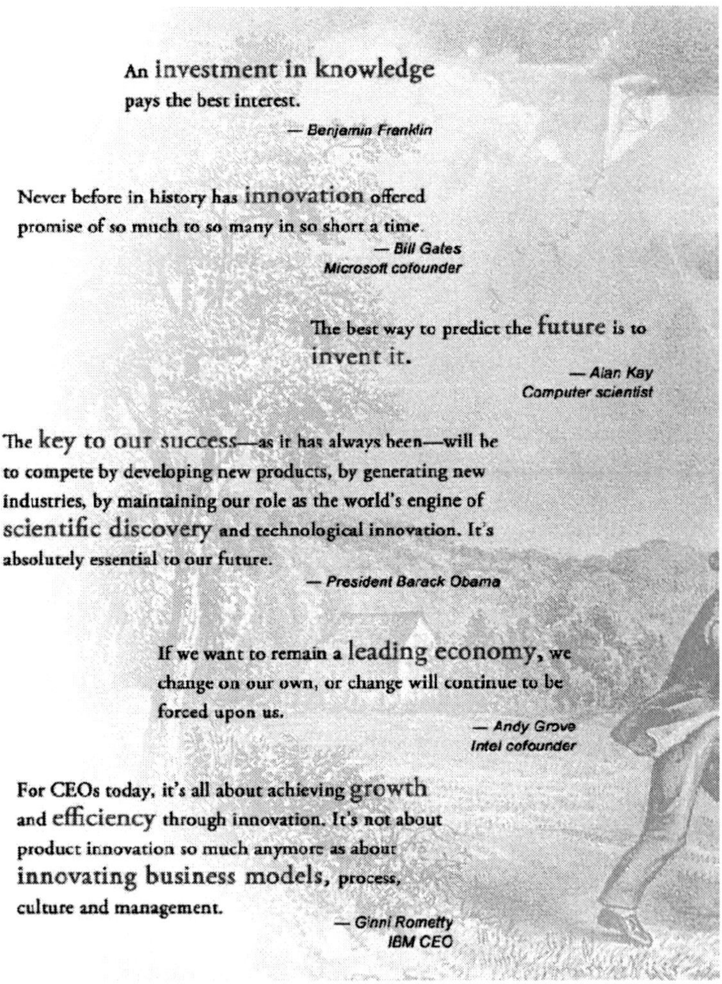

Exceptional Performance

During the 20th century, the pace of innovation was staggering, leading to new industries and companies, such as those in the biotech and information technology fields. Innovation also spurred growth in traditional industries, as businesses fundamentally changed the way they produced and distributed their goods and services.

In the process, the United States became the world's most innovative, most educated, and most competitive nation. Since 1980, the United States made up between 20 and 25 of the world's economy while having only about 5 percent of the world's population.[1] The exceptional economic performance of the United States helped to improve the lives of its citizens, particularly during the decades after World War II. Between 1950 and 2000, incomes soared, with real disposable personal income per capita increasing 213 percent, from $9,240 to $28,899.[2] The U.S. economy created millions of new jobs, many in new firms and industries.

These economic gains were coupled with gains in other areas. The United States provided electricity and phone service throughout the country, built the Interstate Highway System, provided clean water to hundreds of millions, put men on the moon, developed the Internet, and decoded the human genome. (Box 1.1 describes in more detail the construction of the Interstate Highway System.) Advances in medical science helped propel significant increases in life expectancy in the United States. Life spans, as measured at birth, rose from 47.3 years in 1900 to 77.9 in 2007.[3] Advances in agricultural science increased the productivity of our farms by 150 percent between 1948 and 2008.[4]

The United States has a strong tradition of scientific advancement; about 40 percent of Nobel Prizes have been awarded to U.S. citizens[5] and a 2011 study placed 40 percent of the world's 100 most innovative companies in the United States.[6]

Box 1.1. The Interstate Highway System

Perhaps no other infrastructure investment in the United States so transformed society as the Interstate Highway System. The Federal-Aid Highway Act of 1956 was signed into law by President Dwight D. Eisenhower, and it authorized $25 billion for the construction of 41,000 miles of highways (see the map below), at the time making it the largest public works program in American history. As of 1991, construction cost $128.9 billion for about 43,000 of the system's miles. About 90 percent of the funding came from the Federal government, generally from revenue raised by taxes on motor fuel, and about 10 percent from the states.

Today, after more than 50 years, there are more than 46,000 miles in the Inter- state Highway System. It is often touted as one of the greatest public works in- vestments in the nation's history. The Interstate Highway System replaced a lower capacity, lower speed, less safe, and more expensive (per mile of travel) highway system.

The system thus allowed regions and localities that were not part of the nation's economy to become integrated and open to new economic opportunities.[1]

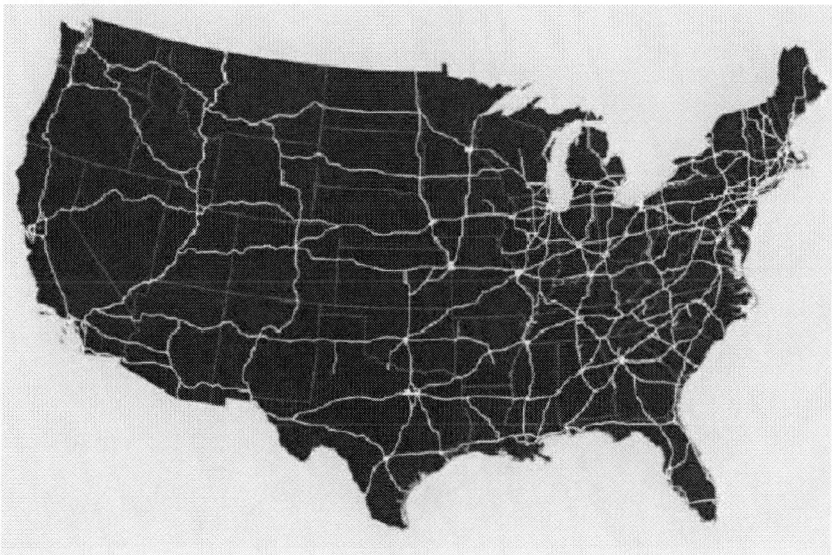

Map of the United States Interstate Highway System.

[1] Source: mappery.com/maps/United-States-Interstate-Map.

States across the country contributed to and benefitted from the U.S. economic strength during the 20th century. From 1963 to 2008, state-level real income per person increased in every state by at least 79 percent, and some states experienced gains of close to 300 percent. A total of 34 states (including the District of Columbia, which had the highest increase) saw growth of more than 150 percent (see figure 1.1).

Successful, world-class companies are located in virtually every state in the United States; 39 states are home to at least one Fortune 500 company. Within states, and across state boundaries, regional innovation clusters arose. Silicon Valley became the world's information technology (IT) epicenter, but other areas also contributed significantly to the IT revolution, including the regional industry clusters in Texas, Washington State, Massachusetts, Georgia, North Carolina, Virginia, and Michigan ("Automation Alley", in Southeast

Michigan). Major medical advances have been made in many states, including Alabama, California, and Pennsylvania.

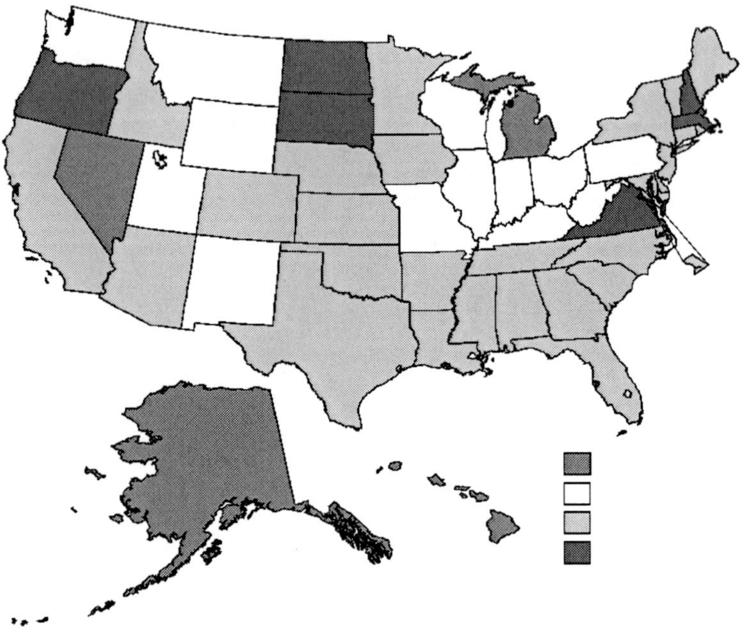

Source: Bureau of Economic Analysis, Economics and Statistics Administration calculations. Note: Percent change in real Gross Domestic Product per capita by state, 1963 (first available year of data) to 2008.

Figure 1.1. Growth in Real GDP per Capita by State, 1963–2008.

Over time, Americans came to take these economic advances for granted and expected these trends to continue into the future. This economic progress fueled belief in an important facet of the American dream—the expectation that our children's quality of life would be better than our own.

Alarms

While the United States exited the 20th century as the undisputed economic and innovation leader, the competitive landscape was shifting.[7] As the economies of more countries around the world grew and developed, these countries became stronger competitors to the United States. Though there are benefits to the United States from these changes, alarms are being raised about these trends and there is also growing angst that the United States is no longer competing as strongly on the world economic stage. One recent poll found that 47 percent of Americans "strongly agree" and 43 percent "somewhat agree" with the statement that the United States is in danger of losing its global competitive edge in innovation.[8] Another survey found that 71 percent of Americans believe that our high schools are falling short when it comes to preparing students for science and engineering jobs and 76 percent believe that if the next generation does not work to improve its science and math skills, it risks becoming the first one that is worse off than its parents' generation.[9]

Alarm 1: Jobs

The United States' ability to create jobs has deteriorated during the past decade. Employment increased at an annual rate of just 0.6 percent between the February 2001 and January 2008 employment peaks (figure 1.2). This rate is one-third as fast as the 1.8 annual rate of employment growth between the June 1990 and February 2001 employment peaks. A recent study by McKinsey Global Institute found that the United States has been experiencing increasingly lengthy jobless recoveries: "it took roughly 6 months for employment to recover to its prerecession level after each postwar recession through the 1980s, but it took 15 months after the 1990–91 recession and 39 months after the 2001 recession."[10]

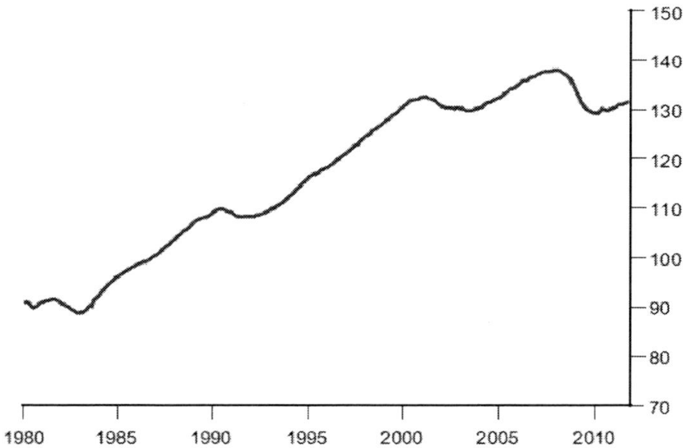

Source: Bureau of Labor Statistics.
Note: In millions of jobs.

Figure 1.2. Nonfarm Payroll Employment, 1980–2011.

Alarm 2: Wages and the Middle Class

The middle class in the United States has struggled as incomes and wages have generally stagnated. One commonly referenced measure of the financial well-being of the middle class is real median household income; that is, the income of households in the middle of the income distribution after adjusting for inflation. From 1980 to its peak in 1999, real median household income increased about 20 percent (see figure 1.3). Since that peak, real median household income has stalled, and even before the Great Recession, real median household income fell from $53,252 in 1999 to $52,823 in 2007 (in 2010 dollars). Individuals at the very top of the income distribution have fared better during this time than others; one study found that between 1993 and 2008, income grew almost 4 percent per year for those with incomes in the top 1 percent of the income distribution.[11]

Source: Household Median Income from the U.S. Census Bureau, Current Population Survey, Annual Social and Economic Supplements.
Note: In 2010 dollars.

Figure 1.3. U.S. Median Household Income, 1977–2009.

The lack of income growth echoes the lack of earnings growth workers have experienced over recent decades. With few exceptions (such as the second half of the 1990s), the typical American worker has experienced long stretches of flat or even declining earnings for full-time work, despite an incredible rise in his or her productivity. Between the fourth quarter 1979 and the fourth quarter 2010 (that is, essentially over the length of a generation), real median weekly earnings of full-time wage and salary workers edged up just 4.9 percent, while workers' productivity increased 90.9 percent. Reasons offered for these wage trends include the decline in the fraction of workers covered by collective bargaining, increased international competition, technological change, immigration, and minimum wages, among others.[12] Regardless of the reasons, this stagnation makes it impossible for many Americans to increase their financial standard of living and feeds the perception that the next generation will be no better off than its parents' generation.

Alarm 3: Manufacturing

These employment and wage trends also roughly coincide with the increased pressure from abroad faced by the U.S. manufacturing sector (though the manufacturing sector has increasingly relied on foreign markets). The manufactured goods trade balance has worsened. In 2010, the trade deficit in manufactured goods was $565.4 billion and is on track to exceed that amount in 2011, even with strong export growth.[13]

Source: U.S. Census Bureau, Foreign Trade Division.
Note: Billions of dollars. In nominal dollars.

Figure 1.4. Advanced Technology Products Trade Balance, 1990–2010.

The United States continues to lose ground in key manufacturing sectors, including those sectors that are likely to drive our economy in the future. The United States ran a trade surplus in "advanced technology products," which includes biotechnology products, computers, semiconductors, and robotics, until 2002 (see figure 1.4).[14] In 2010, however, the United States ran an $81 billion in this critically important sector.[15]

Alarm 4: Innovation

After reviewing 16 key indicators—such as the number of scientists and engineers, corporate and government R&D, venture capital, productivity, and trade performance—the July 2011 *Atlantic Century* report indicated that the United States had made little or no progress in its competitiveness since 1999 and now ranks fourth in innovation-based competitiveness.[16] A report from 2005, *Rising Above the Gathering Storm*—a volume authored by a committee convened in 2005 by the National Academy of Sciences—expressed deep concern that the scientific and technological building blocks critical to the economic leadership of the United States were eroding at a time when many other nations were actively laying strong foundations in these same areas.[17] In their 2010 follow-up report, that same committee unanimously stated that "our nation's outlook has not improved but rather has worsened."[18]

Alarm 5: Education

The United States is struggling to prepare U.S. students in math and science. In 2009, U.S. 15-year-olds had an average score of 487 on the mathematics literacy scale, which was lower than the OECD average score of 496 (see figure 1.5). Seventeen OECD countries ranked above the United States in math, and some 11 other countries had scores that were not significantly different from the U.S. math score. Additionally, science and reading scores were only average and on an earlier assessment of student problem solving ability (2003 Program for International Student Assessment (PISA)), U.S. students scored behind most of the other developed nations in the world.[19]

Alarm 6: Infrastructure

Delays at airports, time lost in traffic jams, bridges in need of repair, and ports that cannot handle the newest ships exemplify how traditional infrastructure in the United States has failed to keep pace with its growing population. The result is higher costs for businesses and inconvenience for all. Digital infrastructure, though stellar in some respects, has not yet reached large portions of our population, making it difficult for them to participate in the 21st century economy. Large and disturbing differences in broadband adoption still persist by income, race and ethnicity, and education.

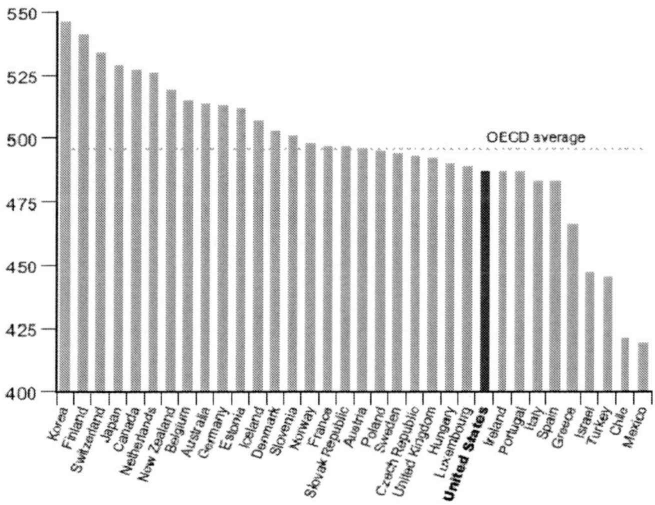

Source: OECD, PISA 2009 database.

Figure 1.5. U.S. Math Test Scores for 8th Graders Remain Below OECD Averages.

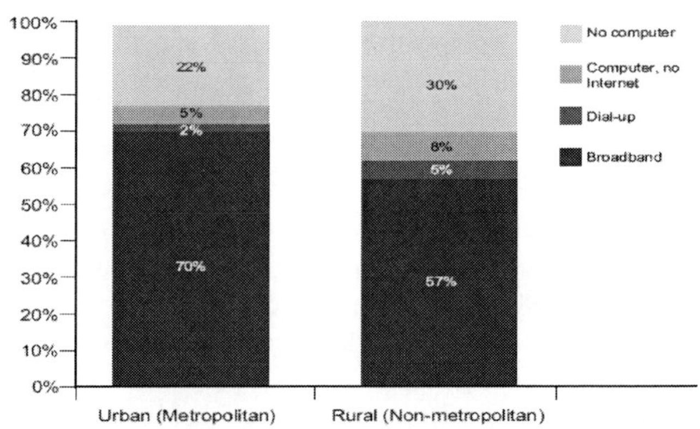

Source: Economics and Statistics Administration and National Telecommunciations and Information Administration. *Exploring the Digital Nation: Computer and Internet Use at Home*. Washington, D.C: U.S. Department of Commerce, November 2011.

Figure 1.6. Computer and Internet Use by Urban and Rural Location, 2010.

Also, some communities are disadvantaged with respect to broadband access and use. For example, those living in urban areas were much more likely to have access to broadband Internet connections relative to rural consumers (see figure 1.6).

Addressing the Alarms

> So, yes, the world has changed. The competition for jobs is real. But this shouldn't discourage us. It should challenge us. Remember—for all the hits we've taken these last few years, for all the naysayers predicting our decline, America still has the largest, most prosperous economy in the world. No workers—no workers are more productive than ours. No country has more successful companies, or grants more patents to inventors and entrepreneurs. We're the home to the world's best colleges and universities, where more students come to study than any place on Earth.
>
> *— President Barack Obama, State of the Union Address, January 2011*

Tough problems need to be tackled in order for the United States to improve its competitiveness and increase good-paying jobs. To address these issues, Chapter 2, "Keys to Innovation, Competitiveness, and Jobs" delves into what made the economy competitive in the past and demonstrates that the Federal government played a key role in research, education, and infrastructure, three components that greatly contributed to the economic vitality of the United States in the 20th century. Put another way, the government (Federal, state, and local) made investments into the building blocks of our economic growth, and these investments allowed the private sector to flourish. Chapters 3, 4, and 5 go into greater detail for each of these broad areas (research, education, and infrastructure), discussing the challenges faced in each and proposed policies to keep the United States at the innovation and competitiveness frontier.

In addition to these key areas, there are other avenues by which the competitiveness of the United States can be increased. For instance, a strong manufacturing sector is crucial, as this sector conducts the majority of industrial research and development, and there are strong links between the location of production and the location of research activity.[20] Given the importance of manufacturing and some of its unique properties, Chapter 6 focuses on manufacturing solely.

Although improving research, education, infrastructure, and the manufacturing sector are essential to increasing innovation and competitiveness, many other factors also contribute to economic success. Perhaps chief among them is ensuring that both established firms and entrepreneurs in the private sector have the best possible environment in which to innovate. Chapter 7 touches upon some of the additional areas where the government can assist the private sector, including aid to regional clusters, promoting entrepreneurship, creating an effective intellectual property regime (domestically and abroad), and reforming corporate taxes.

This report touches upon some of the key policy areas needed to make the U.S. economy more innovative and competitive. Drawing upon the other chapters in this report, the last chapter, "Moving Forward", highlights 10 areas that deserve special attention. Implementing these recommendations will better prepare the United States to meet the economic challenges of the 21st century and provide a better future for our children.

References

Atkinson, Robert D., and Scott M. Andes. 2009. *The Atlantic Century: Benchmarking EU and U.S. Innovation and Competitiveness.* The Information Technology & Innovation Foundation; www.itif.org /publications/atlanticcentury-benchmarking-eu-and-us-innovation-and-competitiveness.

Atkinson, Anthony B., Thomas Piketty, and Emmanuel Saez. 2011. Top Incomes in the Long Run of History." *Journal of Economic Literature 49, no 1: 3–71.* 2011.

Bureau of Labor Statistics. Establishment Data. Table B–1. Employees on nonfarm payrolls by industry sector and selected industry detail. www.bls.gov/webapps/legacy/cesbtab1.htm

Charlton Research Company for Research!America. *Your Congress – Your Health Survey*, March 2011. www.your congressyourhealth.org /where_ america.php?qq=509

Delgado, Mercedes, Michael E. Porter, and Scott Stern. 2011., *Clusters, Convergence, and Economic Performance.* March; www.isc.hbs.edu/pdf /DPS_Clusters_Performance_2011-0311.pdf.

Economic Research Service. 2011. *Agricultural Productivity in the United States: Overview.* Washington, DC: U.S. Department of Agriculture, May; www.ers.usda.gov/Data/AgProductivity/.

Fleischman, H.L., Hopstock, P.J., Pelczar, M.P., and Shelley, B.E. 2010. *Highlights From PISA 2009: Performance of U.S. 15-Year-Old Students in Reading, Mathematics, and Science Literacy in an International Context* (NCES 2011–004). U.S. Department of Education, National Center for Education Statistics. Washington, DC: U.S. Government Printing Office.

International Monetary Fund. 2011. World Economic Outlook Database. September 2011; www.imf.org/external/pubs/ft/weo/2011/02/weodata /index.aspx.

Jensen, J. Bradford. 2011. *Global Trade in Services: Fear, Facts, and Offshoring.* Washington, DC: Peterson Institute for International Economics.

Manyika, James, Susan Lund, Byron Auguste, Lenny Mendonca, Tim Welsh and Sreenivas Ramiswamy. 2011. *An Economy that Works: Job Creation and America's Future.* McKinsey Global Institute, 1; www.mckinsey. com/mgi/ publications/us_jobs/pdfs/MGI_us_jobs_full_report.pdf.

National Academy of Sciences. 2010. *Rising Above the Gathering Storm, Revisited: Rapidly Approaching Category 5.* Washington, DC: National Academies Press.

National Academy of Sciences. 2007. *Rising Above the Gathering Storm: Energizing and Employing America for a Brighter Economic Future.* Washington, DC: National Academies Press; www.nap.edu/catalog /11463.html.

National Center for Education Statistics, Mathematics Age 15, 2009 OECD PISA Data, International Data Explorer, nces.ed.gov/surveys /international /ide/.

National Center for Health Statistics. 2011. "Table 22. Life expectancy at birth, at 65 years of age, and at 75 years of age, by race and sex, 1900–2007" in *Health, United States, 2010: With Special Feature on Death and Dying.* Washington, DC: U.S. Government Printing Office; www.cdc.gov /nchs/data/hus/hus10.pdf#022.

Nobel Prize Facts. Accessed on November 30, 2011; www.nobelprize.org /nobel_prizes/nobelprize_facts.html.

Peter D. Hart Research Associates, Inc. and The Winston Group. 2006. *Keeping Our Edge: Americans Speak on Education and Competitiveness.*

2006. Washington, DC: Hart/Winston; www.ets.org/Media /Education_ Topics/ pdf/HW_KeepingOurEdge2006.pdf.

Piketty, Thomas, and Emmanuel Saez. 2003. "Income Inequality in the United States, 1913–1998." *Quarterly Journal of Economics* 118 (February): 1–41.

Sachdev, Nikhil. 2007. "An Examination of the Wage Productivity Gap." Working Paper. Stanford University. Thomson Reuters. 2011. *Top 100 Global Innovators;* www.top100innovators.com/home.

United States Nobel Prize Winners. Accessed on November 30, 2011; www.jinfo.org/US_Nobel_Prizes.html. U.S. Bureau of Economic Analysis. 2010. *Survey of Current Businessess*. Accessed 14 November 2011; April.

U.S. Census Bureau, Foreign Trade Division. Accessed November 29, 2011. "Advanced Technology Product Definitions" in *Foreign Trade Statistics;* www.census.gov/foreign-trade/reference/glossary/a/atp.html# general.

KEYS TO INNOVATION, COMPETITIVENESS, AND JOBS

Innovation is a key driver of competitiveness, job growth, and a higher standard of living for future generations. To improve the competitiveness of the United States, it is instructive to examine the factors that previously unleashed the tremendous innovative potential of the private sector. The list of those factors is long, and it is not surprising that the recipe for successful innovation and competitiveness is complex, evolving, and differs by product and industry. To address the question of what made the United States innovative and competitive in the past, and also what will make the United States innovative and competitive in the future, this report primarily focuses on three important factors that formed the foundation of a strong innovative environment: support for research, education, and infrastructure.

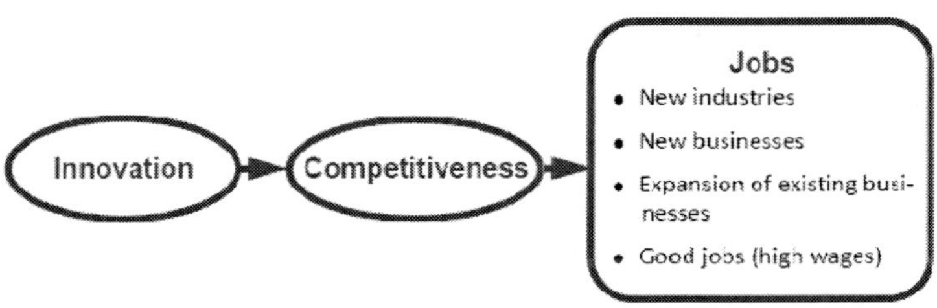

A common thread between these three elements is that they are areas where government has made, and should continue to make, significant investments. In all three of these areas, investment has a social return that exceeds the return to any one company or person. Basic research often has many applications, beyond those which motivated the initial research. A more educated workforce means not just more income for those who attend school longer, but also means greater productivity in business and a more effective citizenry. Improved infrastructure provides a benefit for the greater good and facilitates productivity. Because of these broader benefits, private investment is often too low since private investors cannot capture the broader social returns. As a result, almost all governments in developed countries fund investment in these areas.

Concepts and Definitions

Before delving into these three areas, it is important to take a step back and define several terms. The COMPETES Reauthorization Act directs the Department of Commerce to "complete a comprehensive study of the economic competitiveness and innovative capacity of the United States." It is somewhat ironic, therefore, that the importance of "innovation" and "competitiveness" are matched by the lack of commonly accepted definitions and empirical measures over time and across countries. Beginning with "innovation," a 2008 Advisory Committee report to the Secretary of Commerce, *Innovation Measurement: Tracking the State of Innovation in the American Economy,* defines it as:

> "The design, invention, development and/or implementation of new or altered products, services, processes, systems, organizational structures, or business models for the purpose of creating new value for customers and financial returns for the firm."[1]

There are two main approaches to measuring innovation.[2] The first is the proxy method, where rather than measuring innovation directly, patents or spending on R&D are tracked as a proxy for the level or rate of change of innovation. Although these proxies can be useful tools for understanding innovation, they are necessarily imperfect measures. For example, many innovations are not patented, and innovative activity occurs even in industries that conduct little formal R&D. The second approach relies on economic accounting where economic growth is explained by factors that are measurable, such as the labor force and its quality. The portion of economic growth that cannot be explained by measurable factors is referred to as "technological change," "innovation," or in economic jargon, "multifactor productivity" or "total factor productivity." Using this second approach, it is estimated that between over one-third to a half of economic growth in the United States can be attributed to "innovation."[3]

Similar to innovation, "competitiveness" has also proved difficult to define and measure. A competitive business is one that is successful in the marketplace—success being measured in various ways such as market share or profitability. As the McKinsey Global Institute states, competitiveness in a sector can be defined as the "capacity to sustain growth through either increasing productivity or expanding employment."[4] Though there is not a common definition of competitiveness at the country level, a widely recognized ranking of this comes from the World Economic Forum (WEF). They define competitiveness as "the set of institutions, policies, and factors that determine the level of productivity of a country."[5] Thus, the concepts of productivity and competitiveness often go hand in hand. In this report, the term "competitiveness" is generally used, but often increasing competitiveness requires increasing productivity, and vice versa.

The competitiveness of a country and the competitiveness of businesses are also closely-linked concepts. Competitive businesses need to innovate; otherwise, they will not be able to grow and remain viable. When countries are competitive—that is, when they have a "set of institutions, policies and factors" that are conducive to productivity growth—then businesses are positioned to grow and be effective competitors against other domestic and foreign firms. According to the WEF, "(t)his requires an environment that is conducive to innovative activity, supported by both the public and the private sectors. In particular, it means sufficient investment in research and development (R&D), especially by the private sector; the presence of high quality scientific research institutions; extensive collaboration in research between universities and industry; and the protection of intellectual property."[6] Given the pace of change in today's global economy, investments to promote innovation deserve more emphasis than at any time in the past.

Ensuring a country is competitive and has sufficient capacity to innovate is also crucial because the number and quality of jobs is strongly dependent on these two concepts. As competitive businesses grow, they hire more workers and they also tend to pay well; a number of studies have shown that highly productive firms pay above-average wages.

- **Innovation leads to new industries.** Over the longer-term, new ideas, products, or discoveries can lead to new industries. Examples include the wireless communications industry (290,000 workers in 2007), software and Internet publishing firms and Internet service providers (500,000 workers), and pharmaceutical firms along with companies in biotechnology research and development services (350,000 workers).
- **Innovation leads to new firms.** Between 1980 and 2007, on average over 500,000 new businesses with employees started each year. These new firms produced an average of 3 million new jobs a year.
- **Competitive and innovative firms expand.** Between 1980 and 2007, existing businesses that grew added roughly 13.3 million jobs a year, which translates into an average employment growth rate of 13.9 percent.
- **Competitive and innovative firms create good jobs.** Wages for workers in innovative and competitive firms tend to be higher than wages elsewhere. For instance, firms that export (that is, firms that successfully compete internationally) have been found to pay significant wage premiums.[7] Similarly, a recent report shows that the science, technology, engineering, and mathematics (STEM) workforce earned about 26 percent more than their counterparts in non-STEM occupations. STEM workers also were less likely to experience joblessness, and STEM job growth over the past 10 years was three times faster than growth in non-STEM jobs.

What Made the United States So Successful in the Past?

Many different factors affect innovation and competitiveness and volumes have been written on the economic history of the United States and, more generally, on innovation. However, there is widespread agreement on at least three factors that contributed greatly to the economic strength of the United States during the last century, factors where the government played an important role: support for research, education, and infrastructure. Given the importance of each of these factors, each receives more in-depth treatment in subsequent chapters. Below is a brief description of how important they were in the past century.

Research

Federally funded R&D has resulted in innovations and discoveries, leading to new companies and entire industries that have made Americans more prosperous, healthier, and safer. For example, the first fully electronic U.S. digital computer—the ENIAC—was funded by the U.S. Federal government. For more on the Federal role in the evolution of the computer (see box 2.1). Federal investments in life sciences have decreased mortality and morbidity rates, driving innovations that are at the cutting edge of fighting heart disease, diabetes, cancer, and HIV/AIDS. For example, "the biopharmaceutical industry draws upon (and complements) an exceptionally large publicly funded basic research effort in the life sciences."[8] The investments in health and medicine at the National Institutes of Health (NIH) continue to contribute heavily to advances in the field, and the work of NIH scientists has produced multiple Nobel Prize winners. Cumulative gains in life expectancy after 1900 were worth over $1.2 million to the representative American in 2000, whereas post-1970 gains added about $3.2 trillion per year to national wealth, equal to about half of GDP. Potential gains from future health improvements are also large; for example, it is estimated that a 1 percent reduction in cancer mortality would be worth $500 billion.[9]

Box 2.1. The ENIAC and the IBM 650: Federally Funded Research and the Birth of an Industry

The ENIAC or Electronic Numerical Integrator And Computer was developed to solve the very specific problem of calculating information related to the proper firing of artillery. The ENIAC was developed in the early 1940s by J. Presper Eckert and John W. Mauchly at the University of Pennsylvania, and was funded by the U.S. Army.[1]

From 1945 to 1955 collaborations between the U.S. military, universities, and the private sector led to at least 19 projects related to the development of computers. This collaborative environment helped drive the explosion in innovation, but the bulk of the funding for this research came from the Federal government, with Federal funds accounting for 59 percent of computer related R&D spending by General Electric, IBM, Sperry Rand, AT&T, Raytheon, RCA, and Computer Control Corporation from 1949 to 1959.[2]

Though the funding for these computers primarily came from the Federal government, companies were able to quickly translate the technological advances into commercial applications. For example, IBM was able to combine the benefits of this Federal R&D with its prowess as an existing office equipment producer to create the IBM 650, that sold 1,800 units in the 1950's making it the most commercially successful computer of that period.

These early Federal investments were undertaken without the commercial applications in mind, yet they provided the foundation for the evolution of the computer industry. Seventy years later, the United States. is still reaping the rewards of these early investments. Today, the lives of nearly every American are impacted in some way by the benefits of advances in computer technology. The basic research investments that led to the creation of the early computer are exactly the type of investments that the United States needs to be making today so that future generations will still be reaping the rewards of today's investments for decades into the future.

David C. Mowery. 2011. "Federal Policy and the Development of Semiconductors, Computer Hardware, and Computer Software: A Policy Model for Climate Change R&D?" *Accelerating Energy Innovation Insights from Multiple Sectors.* Chicago: University of Chicago Press, for the National Bureau of Economic Research; 159–188.

Kenneth Flamm. 1987. *Targeting the Computer: Government Support and International Competition.* Washington, DC: Brookings Institution.

Federal investments in materials and military technology underpin the modern military as well as profitable innovations in the private sector. Advancements in chemicals, such as the spike in the production of synthetic rubber during World War II under the Synthetic Rubber Research Program, have spurred innovations in manufacturing that have directly supported national security. Federal investments in atomic physics in the 1930s and 1950s gave rise to the creation of GPS systems, forever changing the deployment of the military, not to mention our daily travels.[10]

The companies that can trace their roots to federally funded research span a wide variety of industries. In their report *Sparking Innovation: How federally funded university research creates innovation, new companies and jobs,* the Science Coalition identifies over 100 companies that Federally funded research helped launch. To provide a flavor of the wide array of companies included in *Sparking Innovation*, Table 2.1 lists a handful of examples that vary greatly by size, location, industry, and Federal funding source.

Education

At the beginning of the 20th century, America led the world in education, and over the following decades the average level of schooling in the United States increased significantly. Americans born in the 1870s had, on average, less than 8 years of formal education.

For the cohort born in 1910, this average had risen to nearly 10 years. For the cohort born in 1940, this average had risen past 12.[11] For cohorts born between 1876 and 1951, average educational attainment grew steadily by nearly 1 year per decade[12] (see figure 2.1, page 2–8).

By the 1950s, the United States enrolled close to 80 percent of its youth in full time secondary schools.[13] The comparison with industrial Western Europe was stark. Among 18 European nations in the 1950s including France, Italy, and Great Britain, each enrolled less than 30 percent of youth in general education secondary schools; all but one (Sweden) were under 20 percent. When youth in technical schools is added, secondary enrollment in Europe did not surpass 40 percent.[14] This gap extended into higher education.

In the 1950s, American enrollment in higher education was expanding rapidly and America's university attainment rates were far higher than any European country. Many factors contributed to the increased college attainment rates, including the GI Bill and an extensive public university system, especially land-grant schools that had a footprint in every state.

Additionally, the college and university system in the United States contains a disproportionate share of the world's most prestigious universities. For example, according to one set of rankings, in 2011–2012, 18 out of the top 25 universities and 30 out of the top 50 universities were in the United States; the United Kingdom was next with four in the top 25 and seven in the top 50.[15] In addition, the United States is the top destination for students studying abroad.[16]

Infrastructure

Throughout the last century, infrastructure investments, supported by the public sector, have been critical to the increased standard of living and economic growth experienced in the United States. For example, water treatment and distribution systems saved lives and facilitated commerce.

Table 2.1. Companies Created as a Result of Discoveries in Federally Funded University Laboratories 9

Company	Location	Year Started	Employment	Innovation	Federal Funding
Arbor Networks	Chelmsford, MA	2000	125	Network security technologies	DOD, NSF
Audyssey Laboratories	Los Angeles, CA	2002	75	Technology fixes impact of room acoustics on sound reproduction	NSF
Buffalo BioBlower Technologies LLC	Buffalo, NY	2005	8	Air sterilization technology for healthcare, homeland security, battlefields	DOD
Cognex Corporation	Natick, MA	1981	729	Industrial machine vision technology	NSF
CREE, Inc.	Durham, NC	1987	3,168	Semiconductor technology increases efficiency of LED, power, and communications products	DOD
FingerlakesAquaculture, Inc.	Groton, NY	1996	11	Aquafilter for economical, large-scale production of farm-raised fish	USDA
Google	Mountain View, CA	1998	19,835	Internet search technology and Web-based applications	NSF
Image Sensing Systems, Inc.	St. Paul, MN	1984	80	Software for monitoring traffic conditions	DOT
ImagiSonix	Sterling, MA	2006	3	Wireless ultrasound for rural, emergency, military,and disaster settings	DOD
iRobot Corporation	Bedford, MA	1990	538	Robots for military, industrial, and consumer use	DOD, NASA
Molecular Imprints, Inc.	Austin, TX	2001	125	"Step and Flash" nano-lithography makes smaller,faster computer chips	DOD
SenSound, LLC	Detroit, MI	2003	8	Technology pinpoints exact source of noise for use in product design, develop-ment, and manufacturing	NSF
TomoTherapy, Incorporated	Madison, WI	1997	665	Machine targets radiation to cancer cells and limits damage to healthy ones	NIH
Universal Display Corporation	Ewing, NJ	1994	80	Organic LED technology for flat panel displays, lasers, and other light generating devices	DOD, DOE
Webscalers	Binghamton, NY	2002	7	Metasearch engines probe deeper into the Web thantraditional search engines	NSF
Xenogen (acquired by Caliper Life Sciences)	Hopkinton, MA	1994	489	In vivo imaging allows scientists to evaluate drugs by observing their effects in living animals	DOD, NIH

Source: The Science Coalition, Sparking Innovation: How federally funded university research creates innovation, new companies and jobs.

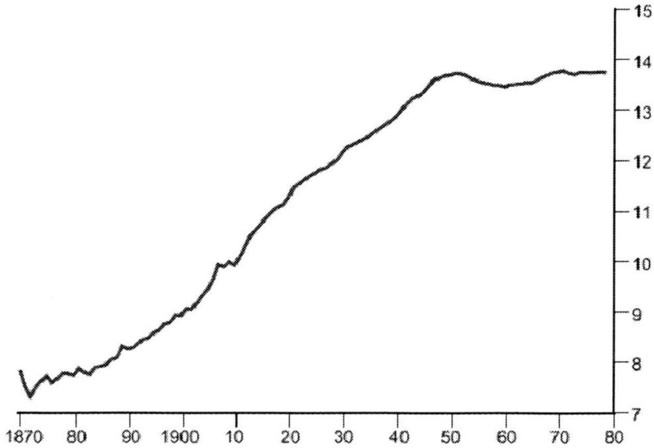

Source: Economics and Statistics Administration (ESA) calculations based on the Integrated Public Use Microdata Series, Minnesota Population Center, University of Minnesota (see http://usa.ipums.org/usa/).

Note: Data for this figure were based on ESA calculations of mean years of education for U.S.-born individuals by birth year for those who were 30 years or older. Because the education variable was coded by category of educational attainment, such as grade levels and higher education levels, it was necessary to transform the data into a continuous variable to calculate a mean. The methodology used to recode the education variable into an estimated number of years of education was based partly on work by Goldin and Katz (2008)."

Figure 2.1. Years of Schooling at Age 30, by Birth Cohorts, 1870–1979.

Early water treatment systems were mostly targeted to protect the public from waterborne diseases, such as typhoid, dysentery, and cholera,[17] but later public water utilities also provided a consistent and dedicated water supply that was important for industrial production and the generation of power, while it also protected the public from environmental contaminants.[18] The interstate highway system, highlighted in box 1.1, was the largest public works project of its time and did more than any other program to connect our country.

Interconnections

Research and development, education, and infrastructure are discussed separately in the chapters that follow, but they are not separate and unique entities. As some commentators have noted, the elements of competitiveness and innovation are less like silos and more like a network or ecosystem.

Changes in one part of the network—say education—ripple through the system satisfying demands for researchers, creating demands for infrastructure, and feeding back into the schools via the creation of demand for new and different skills. U.S. industries, like those discussed in the manufacturing chapter, sit in a critical juncture in this network—creating demand for labor with specific skills and participating integrally in research and in the creation and build out of new infrastructure (see figure 2.2). Thus, although this report addresses innovation and competitiveness topics sequentially in separate chapters, their interconnectedness is a sub-text that the reader should keep in mind.

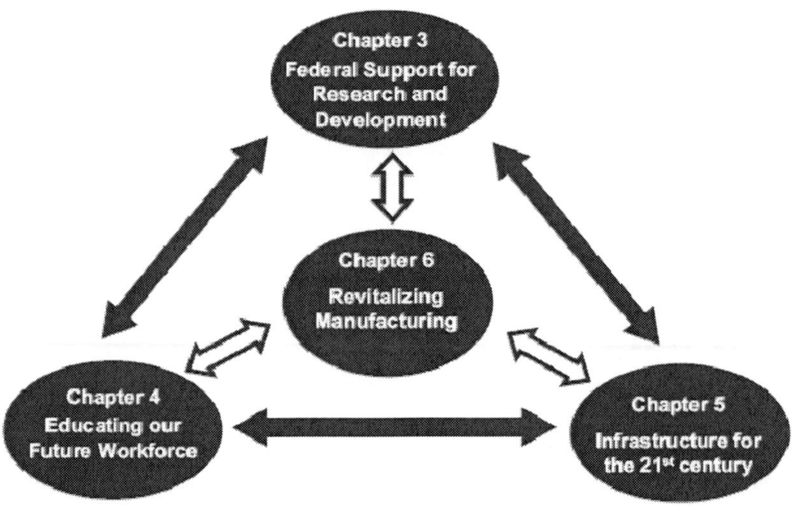

Figure 2.2. The Innovation Ecosystem.

References

Bernard, Andrew B., J. Bradford Jensen, and Peter K. Schott. 2009. "Importers, Exporters and Multinationals: A Portrait of Firms the U.S. that Trade Goods." *Producer Dynamics: New Evidence from Micro Data*, edited by Timothy Dunne, J. Bradford Jensen, and Mark J. Roberts, 513–552. Chicago: University of Chicago Press, for the National Bureau of Economic Research.

Bureau of Labor Statistics. 2011a. "Multifactor Productivity Trends—2009." News Release, March 30; www.bls. gov/news.release/archives/prod3_03302011.pdf.

Bureau of Labor Statistics. 2011b. "Multifactor Productivity Trends in Manufacturing—2009." News Release, August 11; www.bls.gov/news.release/pdf/prod5.pdf.

Bureau of Labor Statistics, 2011c. "Multifactor Productivity Trends for Detailed Industries, 2009." News Release, September 23; www.bls.gov/news.release/pdf/prin3.pdf.

Cockburn, Iain M., Scott Stern, and Jack Zausner. 2011. "Finding the Endless Frontier: Lessons from the Life Sciences Innovation System for Energy R&D." *Accelerating Energy Innovation: Insights from Multiple Sectors,* edited by Rebecca M. Henderson and Richard G. Newell, 113–157. Chicago: University of Chicago, for the National Bureau of Economic Research.

Corrado, Carol A., Charles R. Hulten, and Daniel E. Sichel. 2009. "Intangible Capital and U.S. Economic Growth." The Review of Income and Wealth, 55–3; 661–685.

Committee on Science, Engineering, and Public Policy, National Academy of Sciences, National Academy of Engineering, Institute of Medicine. 1999. *Evaluating Federal Research Programs: Research and the Government Performance and Results Act.* Washington, DC: National Academy Press; www.nap.edu/catalog/6416.html.

Finn, Bernard S. 2002. "Origin of Electrical Power" in *Powering the Past: A Look Back.* National Museum of American History, Washington DC; americanhistory.si.edu/powering/past/prehist.htm.

Goldin, Claudia, and Lawrence F. Katz. 2008. *The Race between Education and Technology.* Cambridge, MA: Harvard University Press.

Jorgenson, D.W. and Zvi Griliches. 1967. "The Explanation of Productivity Change." *The Review of Economic Studies.* Stockholm, Sweden: Institute for International Economic Studies: 34–3; 249–283.

Manyika, James, Lenny Mendonca, Jaana Remes, Stefan Klubmann, Jorg Schubert, Vitaly Klintsov. 2010. *How to Compete and Grow: A Sector Guide to Policy.* McKinsey Global Institute; www.mckinsey.com/Insights/MGI/Research/Productivity_Competitiveness_and_Growth/How_to_compete_and_grow.

Murphy, Kevin M and Robert H. Topel. 2006. "The Value of Health and Longevity." Journal of Political Economy, 114–5; 871–904.

OECD Indicators. 2011. *Education at a Glance 2011*. "Indicator C3: Who studies abroad and where?" Accessed 2 December 2010; www.oecd.org /dataoecd/61/2/48631582.pdf.

Times Higher Education. 2011. "World University Rankings 2011–2012." Thomson Reuters. Accessed 28 December 2011; www.timeshigher education.co.uk/world-university-rankings/.

The Advisory Committee on Measuring Innovation in the 21st Century Economy. 2008. *Innovation Measurement: Tracking the State of Innovation in the American Economy*. U.S. Department of Commerce, Washington DC.

U.S. Environmental Protection Agency. 2000. "The History of Drinking Water Treatment." Fact Sheet; www.epa.gov/safewater/consumer/pdf/hist.pdf.

World Economic Forum. 2011. *The Global Competitiveness Report 2011–2012*; Geneva, Switzerland. reports.weforum.org/globalcompetitiveness.

FEDERAL SUPPORT FOR RESEARCH AND DEVELOPMENT

> "The key to our success—as it has always been—will be to compete by developing new products, by generating new industries, by maintaining our role as the world's engine of scientific discovery and technological innovation. It's absolutely essential to our future."
> — *President Barack Obama, November 17, 2010*

Although it has helped spawn many inventions that, in turn, have led to new firms, new industries, and new jobs, Federal funding of research cannot drive innovation by itself. A healthy private sector must act in partnership with university and research labs to fund the transfer of new technologies to the market, creating new businesses built on innovation. It is also crucial for institutions to encourage research, such as through a strong education system and up-to-date infrastructure. A strong education system ensures there is a workforce with the necessary skills to turn research into practical, market-driven concepts, to make products from those concepts that satisfy consumer preferences and that enhance competition, and to use these products effectively. Infrastructure is necessary to make sure that there is a free flow of ideas, as well as goods and services.[1]

However, the innovative performance of the United States has slipped during the past decade compared to other countries. Looking at a number of measurements of innovation drivers, such as growth in corporate and government research and development (R&D) and the number of scientific and technical degrees and workers, the United States has fallen relative to other countries.[2] Therefore, after describing in more detail the role of R&D in driving innovation and the role of the Federal government in R&D, this chapter concludes with recommendations to help ensure that our country continues to have the innovative capacity it needs to thrive in the 21st century.

The Economic Justification for the Federal Government's Role in Funding Basic Research

Much of the economic growth of recent decades has been driven by innovation.[3] The central role of innovation in economic growth was established through the pioneering work by Abramowitz (1956) and Solow (1957).[4] Increasingly sophisticated models of economic growth in advanced economies have emphasized the crucial role innovation plays.[5] In addition, studies have shown that better training and funding fosters innovation.[6]

Innovation, in turn, is driven in large part by the R&D process, which consists of basic research, applied research and development (for definitions of these items, see Appendix 1). All three of these stages need to thrive in order for innovation to lead to new firms and new jobs.

In 2008, about 60 percent of total public and private R&D spending went to development, with the remaining split about evenly between basic and applied research (see figure 3.1). These proportions have stayed relatively constant over roughly the last 30 years.

Basic economic principles, discussed in more detail in Appendix 2 of this chapter, establish the need for a Federal role in funding R&D, especially in the area of basic research. The knowledge generated by basic research and, to a lesser degree, the application of that knowledge, often shares the characteristics of what is known as a "public good."

A public good has two main characteristics: 1) one person's consumption of that good does not reduce the amount available for others to consume and 2) it is difficult to exclude others from consuming the good. A lighthouse is often considered a classic example of a public good.

Once it is built and operating, everybody sailing in the area will benefit from the lighthouse's operation. It is not possible to sell lighthouse services only to those boat operators that pay for them; their services are available to all who pass.

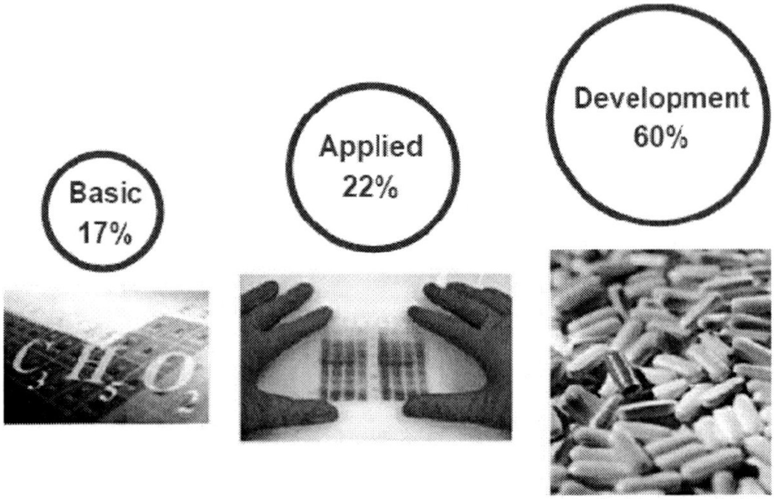

National Science Foundation, Division of Science Resources Statistics. 2010. National Patterns of R&D Resources: 2008 Data Update. NSF 10–314. Arlington, Va. Available at www.nsf.gov/statistics/nsf10314/.

Figure 3.1. The Research Landscape in the United States.

What this means, particularly for basic research, is that it may not be possible for those conducting the research to fully appropriate the benefits from research and innovation. In such cases, the social benefits (those that accrue to society as a whole) from these innovative activities likely exceed the private benefits (those that accrue just to the entity conducting the research). A series of studies show a stark divergence between private and social returns to R&D (see table 3.1). The social return measured in these studies includes the private rate of return plus the change in profit due to R&D spillovers either within an industry or between industries. Because individual researchers cannot recoup the full value of their work, the incentive to produce a socially optimal amount of innovative activity is lacking. This creates a potential role for government to fund innovative activity to raise this activity closer to the social optimum. To accomplish this, the government could directly fund basic research through support of government labs or grants to universities or private research laboratories. Additionally, government policy could increase the returns earned by the private sector on basic research—through policies such as tax credits and a well-functioning patent system—and encourage the private sector to do more basic research.

Table 3.1. Annual Rates of Return on Private R&D Investment

Researcher	Private	Social
Mansfield (1997)	25	56
Sveikauskas (1981)	7–25	50
Scherer (1982, 1984)	29–43	64–147
Bernstein-Nadiri (1991)	15–28	20–110

Source: Center for Strategic and International Studies. Global Innovation/National Competitiveness. Washington, D.C: CSIS, 1996.

Given the public good nature of basic research, it is not surprising that the Federal government plays a stronger role in basic research than in applied research or in the development process. As discussed in more detail below, innovation in the United States has thrived as a result of a research "ecosystem" comprised of three main sectors: the Federal government, the college and university system and the private sector. However, the Federal government, universities, and the private sector all play a different role in terms of the type of research they fund and the type of research they conduct. For example, the Federal government has been the primary funder of basic research, but only conducts a small fraction of all the basic research done in the United States (see figure 3.2). On the other hand, universities conduct about half of the basic research in the United States, but fund a relatively small amount of this research. The private sector, meanwhile, especially the manufacturing sector, funds and conducts most of the applied research and development activity. The total dollars spent by private industry for R&D has been increasing over time and the Federal government must ensure that the university and private sectors have the appropriate incentives to invest in R&D.[7]

The benefits from Federal support of academic research go beyond the development of new and interesting concepts. This is because, when it comes to research and innovation, the Federal government, colleges and universities, and the private sector all are interconnected (see figure 3.3). Federal support of research has positive spillover effects into the other two sectors, and there are also positive spillovers between universities and the private sector. Universities have successfully partnered with the private sector to commercialize technology, with many new companies and jobs resulting from these relationships. An important part of advanced undergraduate- and graduate-level students' education is assisting faculty in federally sponsored research. Such experience prepares students to become part of the nation's science and engineering workforce and to help private firms develop and roll out new technologies.

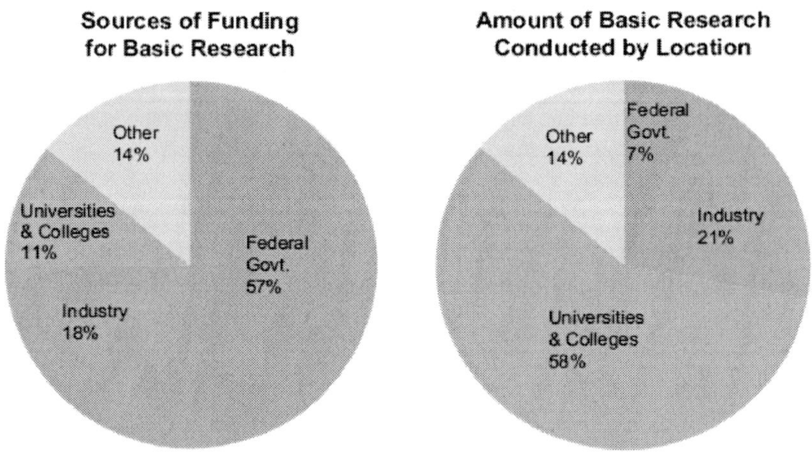

Source: National Science Foundation, Division of Science Resources Statistics. 2010. National Patterns of R&D Resources: 2008 Data Update. NSF 10-314. Arlington, VA. Available at www.nsf.gov/statistics/nsf10314/

Figure 3.2. Sources and Location of Basic Research, 2008.

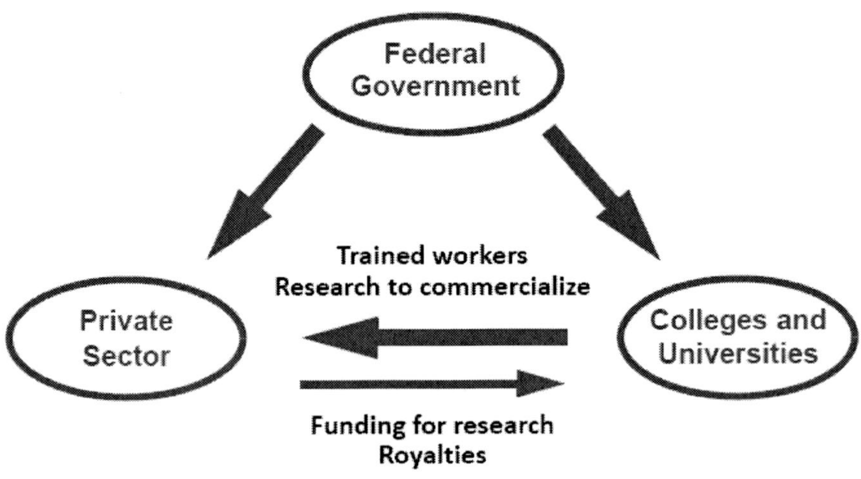

Figure 3.3. The Research "Ecosystem".

A strong research university can also serve as an anchor for the development of a regional innovation cluster (RIC). One way to think about such clusters is that "(r)egional innovation (or industry) clusters are geographic concentrations of interconnected businesses, suppliers, service providers, coordinating intermediaries, and associated institutions like universities or community colleges in a particular field (e.g., information technology in Seattle, aircraft in Wichita, and advanced materials in Northeast Ohio)."[8] New businesses are also generated by RICs; the more than 150 clusters that exist around the country have resulted in increased spin-offs, creating new commercial activity.[9] For example, the *CleanTECH San Diego* cluster initiative launched in 2007 focuses on energy efficiency, renewable energy, transportation and water management. This initiative has generated tremendous startup activity and San Diego now boasts more than 650 clean tech companies, supported by six world class universities and a network of investors. Evidence shows that areas with strong clusters perform better economically than areas without these clusters; they have higher job growth, higher wage growth, more businesses, and a higher rate of patenting[10](see Chapter 7 for more detail on RICs).

The synergies are particularly strong in the manufacturing sector, a sector that has been an important driver of innovation. For example, by training workers and supporting R&D in a number of areas, the manufacturing sector provides a catalyst for product and process innovations for the broader economy. A nation's ability to manufacture products is interconnected with its intellectual and innovative capacity. Many innovative methods and ideas are generated and perfected through the process of making things. Also, the manufacturing sector has tended to play a significant role in the communities where firms are located, as manufacturing plants tend to be large and concentrated, and drive clusters of economic strength within a geographic region. Thus, manufacturing also has proven to be a catalyst for regional clusters, bringing an area benefits such as higher wages.

The Federal government plays a role in facilitating the transfer of research into the marketplace. Recently the President directed Federal agencies to establish measures to monitor the number and the pace of effective technology transfer from Federal labs to nonfederal entities.[11] Agencies are required to develop commercialization plans for their labs that will be monitored by OMB in consultation with OSTP and Commerce. In addition, Commerce will maintain tech transfer metrics to help identify new or creative approaches to accelerate the technology transfer from Federal laboratories to industry.

New initiatives also include efforts to streamline licensing procedures, thereby expanding access to federally-owned inventions, and to use best practices to improve programs directed toward small businesses, such as the Small Business Technology Transfer program. Agencies are also encouraged to launch new programs to support regional innovation clusters by, for example, having their Federal labs share expertise with businesses and by encouraging the location of incubators and research parks near Federal labs. Federal labs and other research facilities will also be encouraged to engage in public-private partnerships that will strengthen commercialization activities in local regions.

The Office of Innovation and Entrepreneurship at the Department of Commerce, in conjunction with its National Advisory Council on Innovation and Entrepreneurship (NACIE) is working to improve commercialization through its i6 Challenge Grants, a competition that funds the best ideas for technology commercialization. In 2011, the i6 Green Challenge followed suit, promoting "Proof of Concept" centers, which support all stages of entrepreneurship, from assisting with feasibility studies and business plan development, to providing access to early-stage capital and mentorship.

The Administration will continue to focus on using prizes to encourage new ways to speed commercialization. Additional initiatives in this area include a joint effort by the Administration, the Association of American Universities, and the Association of Public and Land-grant Universities to encourage university leaders to work more closely with industry, investors, and agencies to increase entrepreneurship, encourage more collaboration between universities and industry, and increase economic development.

The Federal Government: A Key Force Driving Major Innovations

The benefits from Federal R&D support are not just theoretical; whether through funding educational and business organizations or through research in Federal labs, the Federal government has played a crucial role in the development of many key innovations of the mid- to late-20th century. For example, Federal funding, coupled with private industry funding, was critical for the development of the transistor by Bell Labs in the 1950s, the growth of the semiconductor industry, and the birth of Silicon Valley in the 1980s.

The Federal government has also used public-private partnerships to advance markets for key technologies such as the integrated circuit memory chip. For example, the SEMATECH consortium was a partnership created in the late 1980s between the Defense Advanced Research Projects Agency (DARPA) and 14 U.S. based semiconductor manufacturers, including Intel, IBM, Hewlett-Packard, and Texas Instruments.

The Federal government matched the spending put into the venture by SEMATECH member firms and the venture advanced the research needed for the next generation of chips and also funded a test facility to develop prototypes using these new innovations. Most of SEMATECH's members believed they benefited from this arrangement. One member, Intel, invested $17 million in the venture and then reported saving $200 to $300 million as a result of improved yields and greater production efficiencies.[12]

One of the leading examples of how Federal government research support led to significant quality of life improvements in the United States is the development of the Internet. The innovation came about largely because of long-term funding from DARPA in the early 1960s, and then later funding by the National Science Foundation (NSF).[13]

This technology's development relied on basic scientific research that provided evidence it could be used in activities such as packet switching and networking infrastructure.

The financial return from these investments would have been difficult for any single company to capture, and the return could only be seen after many years, making this an ideal candidate for government involvement. Other technologies and businesses related to the Internet also have developed as the result of Federal support, including Google (see box 3.1).

Advances in medical science provide particularly important benefits, given their direct impact on the expected length and quality of life. It has been argued that advances in medical science have probably raised human welfare as much in recent decades as have innovations in all other areas put together.[14]

The National Institutes of Health (NIH), in particular, has been the source of many significant advances in medical science, advances that have improved the well-being of the U.S. population, as well as populations around the world (see box 3.2 for a small sample of the many advances made at NIH over the years and see box 3.3 for a discussion of how Federal support for research led to the creation and expansion of the biotechnology industry.)

Most Federal R&D funding still goes to defense-related activities, while almost half of the Federal non-defense R&D budget goes to NIH (see figure 3.4), with National Aeronautics and Space Administration (NASA) a distant second followed by "All Other" Federal agencies, the Department of Energy, and NSF. Also, Federal spending on basic and applied research has shifted dramatically towards life sciences research, primarily at NIH, over the past two decades. Within just a few years in the late 1990s and early 2000s, NIH spending doubled, while over the same period Federal research expenditures outside of the life sciences grew much less significantly (see figure 3.5). This allocation of research funds contributed to the significant advances achieved through federally supported healthcare-related research.

Box 3.1. How the NSF Seeded Google[1]

In the early stages of developing the Internet, when there were only a few hundred active Web sites, the National Science Foundation (NSF) recognized the need for accessible interfaces for growing online data collections. This led a multi-agency Digital Library Initiative (DLI) that made its first six research awards in 1994. One of those NSF awards supported a Stanford University project led by Professors Hector Garcia-Molina and Terry Winograd.

One of the Stanford graduate students supported by this DLI project was Larry Page. Page was interested in the structure of citations in scientific papers and the way that the citation structure mapped out the knowledge networks in a large and expanding body of scientific literature. He believed the structure mapped out by the linkages across Web sites could facilitate the process of searching for the right site.

Page was joined in this project by another Stanford graduate student, Sergey Brin. Brin's studies at Stanford were supported by an NSF Graduate Student Fellowship. Together, Page and Brin constructed a prototype in their Stanford student offices. The equipment for the prototype, called Backrub, was funded by the DLI project and other industrial contributions. This prototype not only created a text index of linkages across Web pages—it also utilized the structure of linkages across pages to create a web or "tree" of cross-linkages that could facilitate search.

To weight these linkages according to their importance, Page and Brin developed the PageRank method, in which the ranking of a particular Web page depends on the degree to which it is referenced by other frequently referenced Web sites. Page and Brin wrote an early paper on their ideas and tested their algorithm on data from several million Web pages. The results were highly encouraging.

By 1998, Page and Brin obtained funding that allowed them to move their growing operation from Stanford into an off-campus site.

They incorporated Google, Inc. What began as an NSF-funded research project, undertaken by two NSF-supported graduate students, turned into a phenomenon that billions of people around the world use every day.

[1] This account draws heavily from an online summary of *On the Origins of Google*, by David Hart, posted August 17, 2004, on the NSF Web site at www.nsf.gov/discoveries/disc_summ.jsp?cntn_ id=100660.

Box 3.2. The Case for Federal R&D Funding: National Institutes of Health (NIH)

Throughout its long history, NIH has been responsible for numerous research breakthroughs that have contributed greatly to the well-being of the United States and world population. To name just a few:[1]

1968: A Nobel Prize was awarded to Dr. Marshall W. Nirenberg for discovering the key to deciphering the genetic code. He was the first NIH Nobel laureate, and the first Federal employee to receive a Nobel Prize.

1984: In May, scientists uncovered strong evidence that variants of a human cancer virus called HTLV–III are the primary cause of acquired immunodeficiency syndrome (AIDS).

1991: On January 29, NIH scientists treated the first cancer patients with human gene therapy.

1996: The first multicenter trial of bone marrow transplantation in children with sickle cell disease demonstrated that the procedure can provide a cure for young patients that have a matched sibling.

2000: A National Institute of Allergy and Infectious Diseases study showed that a nasal spray flu vaccine not only protected young children against the three strains of influenza for which the vaccine was designed but also a flu strain not covered by the vaccine. It also protected the children against flu-related middle-ear infections.

2000: The international Human Genome Project public consortium—funded by NIH, DOE, and others—assembled a working draft of the sequence of the human genome; it was immediately and freely released to the world.

2005: A long-term, multi-center trial of therapies for high blood pressure found that diuretics work better than newer therapies in treating high blood pressure and reducing the risk of heart disease and should be the first therapy for most patients.

2006: NCI-funded research spanning nearly 2 decades helped lead to the FDA approval for a vaccine to prevent cervical cancer, a disease that claims the lives of nearly 4,000 women each year in the United States.

[1] The full list of accomplishments can be found at www.nih.gov/about/almanac /historical/chronology_of_events.htm.

Box 3.3. The Federal Government and Basic Research: Biotechnology

The creation and success of the biotech industry in the U.S. is due, in great measure, to the Federal government's support of basic research through its funding of the National Institutes of Health (NIH) and the National Science Foundation (NSF).1 From this funding emerged our understanding of the fundamental structure of the human genome and the tools of recombinant DNA technology, without which the biotechnology industry, and the life-saving medicines it is yielding, would not exist as we know it.

The tools of recombinant DNA technology were pioneered by Herbert Boyer, a professor at UC San Francisco, and Stanley Cohen at Stanford University. One of the early pioneers and leaders in this field, Boyer went on to co-found Genentech, together with venture capitalist Bob Swenson. NIH and NSF research funding were therefore instrumental in Genentech's creation. To date, the company employs more than 11,000 people and produces a variety of drugs for asthma, rheumatoid arthritis, and other serious medical ailments.

Protea Biosciences, a company founded in 2001, specializes in the identification of new proteins in the human body. This is important because most pharmaceuticals are small proteins themselves or are small molecules designed to interact with proteins. Today's drugs target fewer than 500 of the estimated 23,000 human protein-coding genes. The technology used to found Protea Biosciences was developed at West Virginia University with support of the WVU Research Corporation and through NIH funding.

[1] Science Coalition, Sparking Economic Growth: How Federally Funded Research Creates Innovation, New Companies and Jobs April 2010.

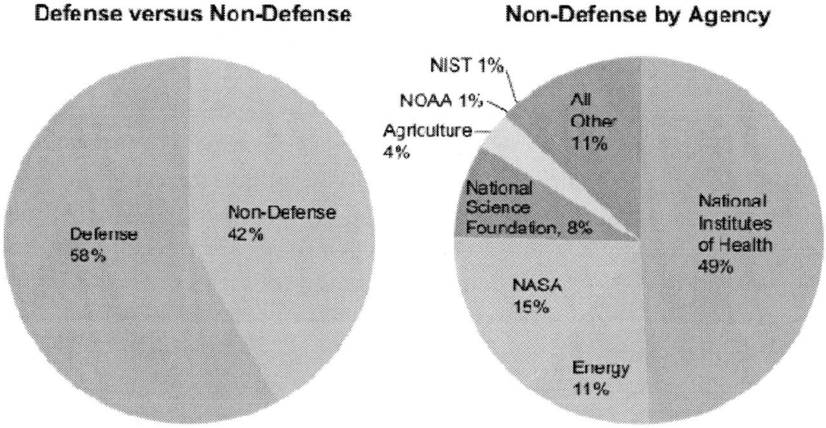

Source: National Science Foundation, Federal R&D Funding by Budget Function: Fiscal Years 2009–11.

Note: Fiscal year 2010 (preliminary).

Figure 3.4. Federal R&D Budget.

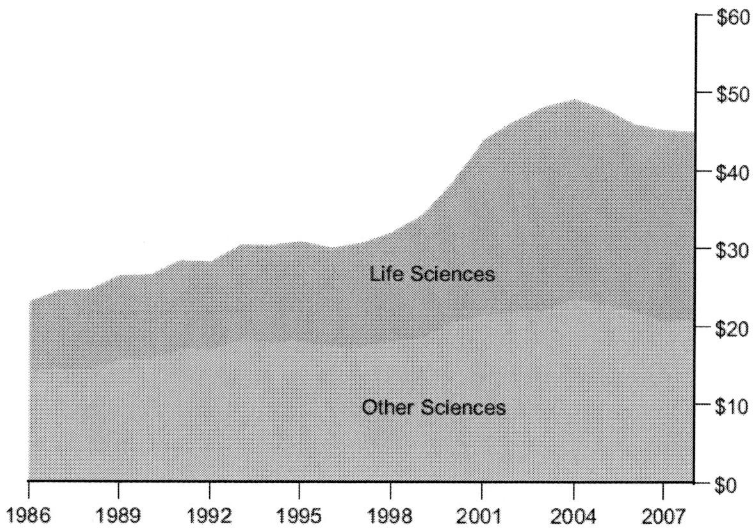

Source: CBO based on NSF, Division of Science Resources Statistics, Federal Funds for R&D: FY 2003, 2004, and 2005 and SEI 2006 (Arlington, VA).

Note: Billions of fiscal year 2000 dollars.

Figure 3.5. Federal Obligations for Research, by Field.

Cracks in the Federal Research Foundation

Although the Federal government's support for research has led to significant innovation, according to the 2011 *Economic Report of the President*, "there are cracks in the foundations of America's growth that need to be addressed. The Nation's innovation system relies largely on the private sector but also depends on critical public inputs. For example, basic scientific breakthroughs in engineering, genetics, chemistry, and many other fields underpin commercial innovation but provide little or no direct profit themselves, so basic scientific research relies heavily on public support."[15]

More specifically, in 1980 the Federal government provided 70.3 percent of all dollars spent on basic research, most of which went to universities and university-based Federal research centers. Since then, the Federal government's share of basic research funding given to all entities has fallen to 57.0 percent and its share of funding of basic research at universities has fallen to about 60 percent, largely due to increased funding from the private sector.[16]

There also has been a slowdown in commercialization of technologies by U.S. universities since 2000. In 1980, Congress passed the Bayh-Dole Act, which gave ownership of the intellectual property to the universities and institutions that created it, even if they used Federal dollars to conduct the research. This was meant to provide a strong incentive for universities to offer useful technology to industry, who would then quickly transform it into products. By the late 1980s, university patenting, licensing of technology to industry, and the proliferation of university-linked startup companies all began to accelerate, reaching especially high growth rates in the late 1990s. However, the pace of these activities slowed starting in 2000, a slowdown that persisted after the brief recession of the early 2000s.[17]

Another area that may be suffering from a lack of sufficient funding for research and innovation is manufacturing, particularly in the area of "advanced technology products" (ATPs). ATPs include goods such as biotechnology products, solar cells, photosensitive diodes, computers, semiconductors, and robotics[18] and it is crucial for our economy that we remain strong in these areas. As Susan Hockfield, president of M.I.T. put it, "(t)o make our economy grow, sell more goods to the world and replenish the work force, we need to restore manufacturing—not the assembly-line jobs of the past, but the high-tech advanced manufacturing of the future."[19] The United States was running a trade surplus in these manufacturing products until 2002, when it lost its advantage in ATPs; in 2010, the United States ran an $81 billion trade deficit in this critically important sector.[20]

Preserving and Extending Federal Support for Science and Industrial R&D in the 21st Century

As the Administration has stated, "there are disturbing signs that America's innovative performance slipped substantially during the past decade. Across a range of innovation metrics—including growth in corporate and government R&D, the number of scientific and technical degrees and workers, access to venture capital, and the creation of new firms—our nation has fallen in global innovation-ranked competitiveness."[21] For example, according to the World Economic Forum, the United States was ranked 7th in the world in its innovative capacity.[22]

To some degree, this is inevitable as other countries become more developed and wealth spreads more equally around the world. However, many countries "recognize that innovation is the key to long-term economic growth and are making pro-innovation investments and adopting pro-innovation policies. Without thoughtful, decisive, and targeted actions, we cannot expect that the industries of the future will emerge and prosper in the United States."[23] Therefore, we are recommending the following policies so that the United States can maintain its position as a world leader in innovation.

Continue to Increase Government Funding for Basic Research

Various documents, including the America COMPETES Reauthorization Act of 2010 which mandated this report, have highlighted the critical importance of the NSF, the Department of Energy's Office of Science (DOE SC) and the National Institute of Standards and Technology (NIST) laboratories in the area of maintaining the United States' leadership role in innovation. These entities need continued support. Also, basic biomedical research such as that done by NIH, also contributes significantly to innovation and deserves continued support.

Sustain Government Funding for Research

In the long run, scientific output will be, to a great extent, a function of the quantity and quality of individuals who are induced to choose science as a career. However, a quality scientific education takes a long time, so rapid increases in public funding in particular fields, followed by sharp cutbacks, can negatively affect the career prospects of young doctorates in the field and discourage younger students who might consider going on to the next level of training (see box 3.4). Stable funding would help ensure that the nation receives the full benefit of its long-run investments in R&D.

Box 3.4. The Changing Nature of Scientific Endeavors

The nature of scientific endeavors has changed greatly and policies to improve R&D need to keep up with these changes. For example, the time required to educate and train new scientists has increased greatly.[1] The body of knowledge a new researcher must absorb has increased and younger scientists must specialize in narrower technical areas. Yet, the solutions to technical problems typically lie outside any one field and scientists must collaborate in teams.[2] Agencies like the National Science Foundation (NSF) and National Institutes of Health (NIH) have recognized these changes, and are actively working to promote team science and young investigators through a number of new funding strategies.

On top of this, talented young people must choose between science, which requires an ever-growing period of education and apprenticeship, and careers in fields such as law and finance, which require a shorter period of education. Thus, it may be necessary to change how young scientists are educated, compensated, and evaluated for research grants to preserve adequate incentives for outstanding young people to enter scientific fields of study.[3]

[1] Benjamin F. Jones "Age and Great Invention," *Review of Economics and Statistics* 92 (February 2010) 1–14.

[2] See Jones (2010).

[3] Former NIH director Elias Zerhouni identified this age trend as the most important challenge for American science and its funding agencies. See Jocelyn Kasier "Zerhouni's Parting Message: Make Room for Young Scientists," *Science* 322 (November 2008), 834–35.

Incentivize and Reward Private Sector R&D Investment with an Enhanced and Extended R&D Tax Credit

Although the Federal government's role in R&D is crucial, private R&D investment remains essential if ideas are to move from university labs and factories to commercialization. Therefore, the Administration has proposed simplifying, enhancing, and extending a corporate R&D tax credit, one that is properly structured so that it awards firms for undertaking additional R&D, not just activity that would have occurred even without the credit. This would reward private industry for undertaking the risks associated with R&D spending, and it would address the reality that private sector inventors often create social benefits that far exceed the private returns to R&D. The

Administration expects this tax credit would provide over $100 billion in benefits to industry over the next decade.

Support Innovative Entrepreneurs

Entrepreneurs and new firms play an essential role in the process of how scientific discoveries are translated. They develop new industries, create jobs, and spur economic growth. The financial crisis and the recession from which we are still recovering disrupted the normal financial and support channels for these entrepreneurs. Passage of the Small Business Jobs Act provided an additional $44 billion in loans through the Small Business Administration and Treasury, and it also provided $12 billion in tax relief to small businesses. The Administration seeks to build on its efforts in this area through its Startup America initiative, which will continue to improve access to capital for start-ups and accelerate commercialization of new technology.

Speed the Movement of Ideas from Basic Science Labs to Commercial Application

The Administration is committed to continue its i6 Green Challenges to develop "Proof of Concept" centers to support all stages of the entrepreneurship process. As venture capitalists often invest in enterprises that are close to marketing a product, researchers can find it difficult to get early-stage funding for their ideas. Proof of Concept centers can help bridge that gap. In September 2011, 6 initial winners of these grants were announced, including the Iowa Innovation Council, whose Proof of Concept center is meant to improve interactions between entrepreneurs, businesses, and universities; accelerate technology transfer; and facilitate company and job creation.[24]

The Administration's Advanced Manufacturing Partnership seeks to identify opportunities for industry, academia, and government to collaborate in order to accelerate the development and deployment of emerging technologies with the potential to transform and reinvigorate advanced manufacturing in the United States.[25] NIH has created a new National Center for Advancing Translational Sciences that will speed the development of new diagnostics, treatments, and cures by building new bridges between the lab and clinic. In addition, the Administration is developing a Bioeconomy Blueprint detailing ways to use biological research innovations to address challenges in areas such as health, food, energy, and the environment. The Blueprint will focus on how to speed up commercialization and open new markets and more workforce training to develop more scientists and engineers.

Unleash a Clean Energy Revolution

New and improved energy technologies will be central to the 21st century global economy, and the Obama Administration is committed to fostering American leadership in this area. These technologies will provide economic growth, create jobs, reduce manufacturing costs, and confront environmental challenges while enhancing energy security. Industrial progress in this area will require a new foundation of fundamental breakthroughs on which it can build. As a part of the vision for doubling America's use of clean energy by 2035, the Administration is also committed to accelerating the deployment of clean energy options that are commercially viable today through such activities as the Renewables Rapid Response Team or the Rapid Response Team for Transmission. The Administration also supports policies, such as a Clean Energy Standard, which provide certainty and guidance for future private sector investment in energy generation.

Accelerate Biotechnology, Nanotechnology, and Advanced Manufacturing R&D

Various advanced technologies are already showing great promise and efforts should be expanded to ensure these technologies reach their full potential. For example, in the area of biotechnology, Federal funding is being provided to increase the number of individual human genomes sequenced from 34 to over 1,800, with the goal of providing insight into the causes and treatments of major diseases and to bring down the cost of sequencing. The National Nanotechnology Initiative is also helping to foster promising developments in the area of nano-technology, "the study and application of extremely small things."[26] Materials that are made at the nanoscale have desirable properties, such as lighter weight, more strength, and greater chemical reactivity. Nanotechnology materials are already used in a wide range of products, such as surface treatments of fabrics to resist wrinkles or staining and high-power rechargeable batteries for cars. Nano-technology has also shown promise in areas such as disease prevention (nano-devices to transport healthy genetic material to cells), self-management interventions (noninvasive detection of glucose levels in diabetic patients), and disease detection (quantum dots to detect cancer cells). The goal of the initiative is to further advances in this field.[27] As mentioned above, the Advanced Manufacturing Partnership will seek to develop and deploy advanced manufacturing processes and technologies to help United States manufacturing continue its outsized contribution to America's economic recovery.

Develop Ways to Measure the Value and Effectiveness of Research Investment

In order to ensure that R&D funding is being spent wisely, it is crucial that meaningful measurement tools are developed to track the effectiveness of this spending. Currently, such measures generally do not exist or are not collected on a regular, systematic basis. One exception to this is the Science and Technology for America's Reinvestment: Measuring the Effect of Research on Innovation, Competitiveness and Science (STAR METRICS). A pilot venture led by NIH, NSF, and OSTP, STAR METRICS will collect data from a number of large research institutions funded by the Federal government to calculate employment effects generated from certain Federal science research funding and investigate ways to measure outputs such as patents, business start-ups, and publications at these institutions.[28] Going forward, additional measures need to be developed and collected on a regular and timely basis.

Appendix 1. Definitions of Relevant Terms

Innovation is the design, invention, development, and/or implementation of new or altered products, services, processes, systems, organizational structures, or business models for the purpose of creating new value for customers and financial returns for the firm.[29]

R&D, also called research and experimental development, comprises creative work undertaken on a systematic basis to increase the stock of knowledge—including knowledge of man, culture, and society—and its use to devise new applications.

Research is the systematic study directed toward fuller scientific knowledge or understanding of the subject studied. Research is classified as either basic or applied according to the objectives of the sponsoring agency.

Basic research is the systematic study directed toward fuller knowledge or understanding of the fundamental aspects of phenomena and of observable facts without specific applications towards processes or products in mind.

Applied research is the systematic study to gain knowledge or understanding necessary to determine the means by which a recognized and specific need may be met.

Development is the systematic application of knowledge or understanding directed toward the production of useful materials, devices, and systems or methods, including design, development, and improvement of prototypes and new processes to meet specific requirements (see www.nsf.gov/statistics /nsb1003/ definitions.htm).

Appendix 2. The Theoretical Underpinnings for a Federal Role in Research Funding

Economic growth results from several factors, but among the most important in recent decades has been innovation, which we will define as the design, invention, and development of new or altered products, services, and processes for the purpose of creating new value for customers and financial returns for the firm.[30] Economists established the central role played by innovation in economic growth in the 1950s, when early empirical efforts to account for growth in U.S. output by measuring labor and capital inputs left the largest part of growth unexplained. Pioneering work by Abramowitz (1956) and Solow (1957) pointed to improvements in technology as constituting the single most important driver of increases in U.S. output per person.[31] In the 1980s and 1990s, increasingly sophisticated efforts by economists to define the growth process in advanced industrial economies placed the process of invention at the center of their models.[32] In addition, studies have shown that improvements in technology are themselves the outcome of deliberate efforts to invent and/or adopt new technology; that is, innovation does not need to be left to its own devices but can be fostered through training and funding.[33]

At least in the long run, efforts to raise per capita income through additional investments in physical capital will run into diminishing returns. But innovation need not be subject to these diminishing returns. Technological progress can, in principle, drive economic growth without limit, thanks to the unique properties of technological knowledge as an economic asset. In addition, innovation is *nonrival*, in the sense that one person can consume it without diminishing the consumption of another party. Thomas Jefferson gave a characteristically poetic expression of this idea when he observed that, "he who receives an idea from me, receives instruction himself without lessening mine; as he who lights his taper at mine, receives light without darkening me."[34] But knowledge also tends to be, at least in part, *nonexcludable*, which means that it is difficult to prevent another party from using the good and deriving benefits.

Because innovation can create knowledge with the attributes of nonrivalry and nonexcludability, it can have some of the classic aspects of a "public good" and may be undersupplied by the market economy. Self-interested agents in a market driven economy will invest only in what they can derive profit from. When the benefits created by an invention cannot be fully appropriated by the inventor, she will create less—perhaps far less—than is socially optimal. We are left with a classic market failure, in which the private value of innovative activity is lower—perhaps far lower—than the social value. That divergence creates a potential role for government intervention to promote innovative activity.

REFERENCES

Abramowitz, Moses. 1956. "Resource and Output Trends in the United States Since 1870." *The American Economic Review* 46 (May): 5–23.

Acemoglu, Daron. 2008. *Introduction to Modern Economic Growth.* Princeton, NJ: Princeton University Press; press.princeton. edu/titles /8764.html.

The Advisory Committee on Measuring Innovation in the 21st Century Economy. 2008. *Innovation Measurement: Tracking the State of Innovation in the American Economy.* Washington, DC: Economics and Statistics Administration, U.S. Department of Commerce; www.esa. doc.gov/sites/default/files/reports/documents/innovation_measurement_ 0108.pdf.

Atkinson, Robert D., and Scott M. Andes. 2009. *The Atlantic Century: Benchmarking EU and U.S. Innovation and Competitiveness.* The Information Technology and Innovation Foundation; www.itif.org /publications/atlanticcentury-benchmarking-eu-and-us-innovation-and-competitiveness.

Arrow, Kenneth J.1962, "The Economic Implications of Learning by Doing," *Review of Economic Studies* 29, no. 3 (June): 155–73.

Bernstein, Jeffrey I., and M. Ishaq Nadiri. 1991. *Product Demand, Cost of Production, Spillovers, and the Social Rate of Return to R&D.* Working Paper no. 3625. Cambridge, Mass.: National Bureau of Economic Research. February.

Burrows, Peter. 1992. "TQM Reality Check: It Works, but It's Not Cheap or Easy." Electronic Business 18, no. 11 (August): 47–52.

Council of Economic Advisers. 2011. "The Foundations of Growth." In *The Economic Report of the President.* Washington, DC: U.S. Government Printing Office, February, 53–79.

Delgado, Mercedes, Michael E. Porter, and Scott Stern. 2011. *Clusters, Convergence, and Economic Performance* (March); www.isc.hbs.edu/pdf /DPS_Clusters_Performance_2011-0311.pdf.

Economic Development Administration. 2011. "Obama Administration Announces $12 Million 16 Green Investment to Promote Clean Energy Innovation and Job Creation." Press Release. September 29; www.eda. gov/NewsEvents/PressReleases/20110929_i6GreenChallenge.xml.

Furman, Jeffrey L., Michael E. Porter, and Scott Stern. 2002. "The Determinants of National Innovative Capacity." *Research Policy* 31, no. 6 (August): 899–933.

Griliches, Zvi. 1979. "Issues in Assessing the Contribution of Research and Development to Productivity Growth." *Bell Journal of Economics* 10, no. 1 (Spring): 92–116.

Griliches, Zvi. 1958. "Research Cost and Social Returns:" Hybrid Corn and Related Innovations." *Journal of Political Economy* 66, no. 5 (October): 419–431.

Griliches, Zvi. 1957. *Hybrid Corn: An Exploration in the Economics of Technological Change. Econometrica* 25, no. 4 (October): 501–522.

Hockfield, Susan. 2011. "Manufacturing a Recovery." *New York Times*, August 29. www.nytimes.com/2011/08/30/ opinion/manufacturing-a-recovery.html.

Irwin, Douglas A., and Peter J. Klenow. 1996., "SEMATECH: Purpose and Performance." In *Proceedings of the National Academy of Sciences* (November 12): 12739–12742. www.ncbi.nlm.nih.gov/pmc/articles /PMC 34130/.

Litan, Robert, Lesa Mitchell, and E. J. Reedy. 2007. "Commercializing University Innovations: Alternative Approaches." In *Innovation Policy and the Economy,* ed. Adam Jaffe, Joshua Lerner, and Scott Stern, 31–57. Chicago: University of Chicago Press, for the National Bureau of Economic Research.

Muro, Mark, and Bruce Katz. 2010. "The New 'Cluster Moment': How Regional Innovation Clusters Can Foster the Next Economy." Metropolitan Policy Program at Brookings Institution, September; www.brookings.edu/~/media/Files/rc/papers/2010/0921_clusters_muro_ katz/0921_clusters_muro_katz.pdf.

National Economic Council, Council of Economic Advisers, and Office of Science and Technology Policy. 2011. *A Strategy for American Innovation: Securing Our Economic Growth and Prosperity.* Washington,

DC, February; www.whitehouse.gov/sites/default/files/uploads/InnovationStrategy.pdf.

National Nanotechnology Initiative. Accessed November 2011. *Benefits and Applications. nano.gov/you/nanotechnology-benefits.*

National Nanotechnology Initiative. Accessed November 2011. *What is Nanotechnology?* nano.gov/nanotech101/what/definition.

National Science Foundation. 2010. *National Patterns of R&D Resources, 2008 Data Update;* www.nsf.gov/ statistics/nsf10314/content.cfm?pub_id=4000&id=2.

Nordhaus, William. 2005. "Irving Fisher and the Contribution of Improved Longevity to Living Standards." *The American Journal of Economics and Sociology* 64, no. 1 (January): 367–92.

Office of Science and Technology Policy, Executive Office of the President. 2011. *'Lab to Market' Initiatives Transforming New Ideas into New Jobs.* Posted on October 28; www.whitehouse.gov/blog/2011/10/28/lab-market-initiatives-transforming-new-ideas-new-jobs.

Romer, Paul. 1990. "Endogenous Technological Change." *Journal of Political Economy* 98 (October): S71–S102.

Romer, Paul. 1986. "Increasing Returns and Long-Run Growth." *Journal of Political Economy* 94, no. 5 (October): 1002–1037.

Scherer, Frederic M. 1984. "Using Linked Patent and R&D Data to Measure Interindustry Technology Flows." Chapter 20 in *R&D, Patents, and Productivity,* Zvi Griliches, ed. Chicago: University of Chicago Press.

Scherer, Frederic M. 1982. "Inter-Industry Technology Flows and Productivity Growth." *Review of Economics and Statistics*, vol. 64, no. 4 (November): 627–634.

Solow, Robert M. 1957. "Technical Change and the Aggregate Production Function." *Review of Economics and Statistics* 39 (August): 312–320.

Sveikauskas, Leo. 1981. "Technology Inputs and Multifactor Productivity Growth." *Review of Economics and Statistics*. vol. 63, no. 2 (May): 275–282.

"Thomas Jefferson to Isaac McPherson." 1813. In *The Writings of Thomas Jefferson* [volume 3], edited by Andrew A. Lipscomb and Albert Ellery Bergh, 333–335. Washington, DC: Thomas Jefferson Memorial Association, 1905; press-pubs.uchicago.edu/founders/documents/a1_8_8s12.html.

U.S. Census Bureau, Foreign Trade Division. 2010. "Advanced Technology Product Definitions." In *Foreign Trade Statistics*; www.census.gov/foreigntrade/reference/glossary/a/atp.html#general.

U.S. Department of Health and Human Services, National Institutes of Health, "STAR METRICS: New Way to Measure the Impact of Federally Funded Research," Press Release, June 1, 2010; www.nih.gov/news/health/jun 2010/od-01.htm.

White House, Office of the Press Secretary. 2011a. "Accelerating Technology Transfer and Commercialization of Federal Research in Support of High-Growth Businesses." Presidential Memorandum. October 28; www.white house.gov/the-press-office/2011/10/28/presidential-memorandum-accele rating-technology-transferand-commerciali.

White House, Office of the Press Secretary. 2011b. "President Obama Launches Advanced Manufacturing Partnership." News Release. June 24; www.whitehouse.gov/the-press-office/2011/06/24/president-obama-launchesadvanced-manufacturing-partnership.

World Economic Forum. 2011. *The Global Competitiveness Report, 2011–2012.* Geneva, Switzerland: Centre for Global Competitiveness and Performance; reports.weforum.org/global-competitiveness-2011-2012.

EDUCATING OUR WORKFORCE

> "If we want to win the future—if we want innovation to produce jobs in America and not overseas—then we also have to win the race to educate our kids."
>
> *—President Barack Obama, State of the Union Address, January 25, 2011*

Education is a key element for promoting economic growth and increasing the innovative capacity of a firm or a country. Economic growth "closely depends on the synergies between new knowledge and human capital, which is why large increases in education and training have accompanied major advances in technological knowledge in all countries that have achieved significant economic growth."[1] Our nation's education system underpins the United States' rise to the position of richest nation on the planet in the last century.[2] However, we must recognize and address cracks in this building block of American innovation, lest we fall behind countries that have placed a higher priority on developing a skilled workforce.

It is not sufficient in today's global economy for a nation to have a generally skilled and educated workforce. Increasingly, the specific skills embodied in science, technology, engineering, and mathematics (STEM) education fuel the innovative processes that are especially valuable to our economy. These skills are sought by companies across the economy as they look to expand their work-forces. These STEM skills are not only important for those working towards advanced degrees. All levels of the education system should incorporate the critical thinking and other skills that are the hallmark of STEM education.[3]

This chapter compares the United States to other nations on the dimensions of access to education and training and academic outcomes, with a particular focus on STEM. Furthermore, it outlines the diverse and critical role of the Federal government in building a skilled and competitive workforce. Areas to be addressed are summarized below:

- The United States must sustain the quality of its post-secondary education system, which is the top destination for students from abroad, while also removing barriers that have limited the post-secondary participation and performance of U.S. students. It is essential that the United States equip future and current workers with the skills needed to compete in a global labor market.
- Given the importance of the role played by technological progress and innovation in promoting economic growth, investment in STEM education is especially important. Yet the United States is falling behind in this area at all education levels, and addressing this shortcoming is needed if we are to continue to produce not only a workforce with the technical skills needed to fill current job openings, but persons with the unique blend of technical expertise and entrepreneurial spirit who will create the products and industries of the future.

Education is a complex and multifaceted process that spans pre-school through life-long learning and involves policy issues ranging from affordability and technology, to questions of support for higher education, classroom size, equal access, and teacher compensation. This chapter primarily and narrowly focuses its attention to STEM because of the strong link between STEM skills, STEM occupations, and innovation. However, our narrow attention to STEM in no way implies that other aspects of education policy are not important in making our country more innovative and competitive. Indeed, our attention to STEM should be viewed as only one example of an area where concern has been raised about the nation's performance relative to other countries in the world.

The STEM Workforce Is Expanding

The STEM workforce is typically defined as the set of professional and technical support occupations in the fields of computer science and mathematics, engineering, and life and physical sciences. In 2010, there were 7.6 million STEM workers in the United States, representing about 1 in 18 workers. Computer and math occupations account for close to half of STEM employment, followed by engineering with 32 percent of STEM jobs, physical and life sciences with 13 percent, and STEM management jobs with 9 percent. Over the past 10 years, growth in STEM jobs (7.9 percent) was three times as fast as growth in non-STEM jobs (2.6 percent). Looking ahead over the coming years, STEM employment is expected to continue to grow at a faster rate (see figure 4.1).

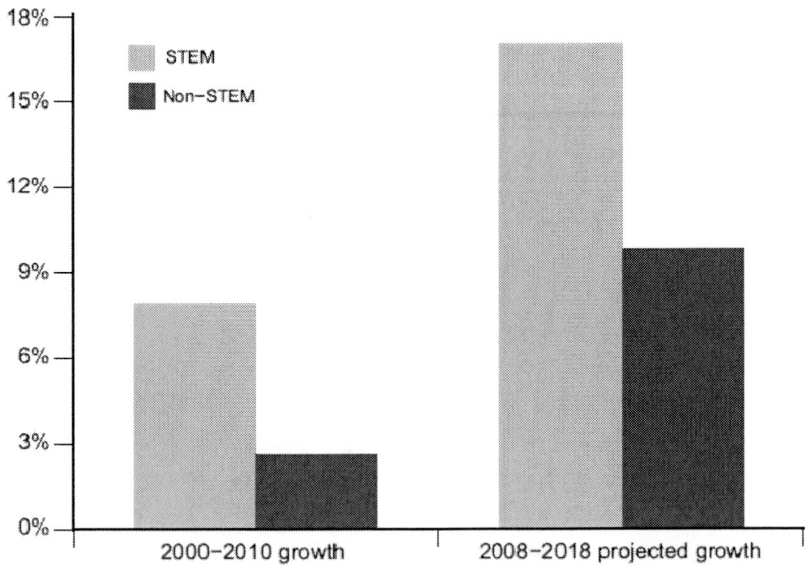

Source: Economics and Statistics Administration calculations using Current Population Survey public-use microd at a and estimates from the Bureau of Labor Statistics' Employment Projections Program.

Figure 4.1. Recent and Projected Growth in STEM and Non-STEM Employment.

STEM workers fill our nation's research and development facilities and drive our nation's innovation and competitiveness by generating new ideas, new companies, and new industries. Not surprisingly, more than three-fourths of the most celebrated inventors and entrepreneurs since 1800 had degrees in engineering, physics, chemistry, computer science, or medicine.[4]

Commensurate with their importance in driving economic productivity and growth, workers in STEM fields earn more on average than workers in other fields. As a result, providing more students with the skills to work in STEM fields is crucial both to the nation's economic future and to improving the incomes of our workers. STEM workers enjoy large earnings premiums over non-STEM workers. For example, in 2010, the STEM premium earned by workers with a bachelor's degree was 27 percent, and for workers with a graduate degree, it was 12 percent[5] (see table 4.1). STEM workers are also less likely to experience joblessness than their non-STEM counterparts.

Table 4.1. Average Hourly Earnings of Workers in STEM Occupations, 2010

Education	STEM	Non-STEM	Difference
High school diploma or less	$24.82	$15.55	59.6%
Some college or associate degree	$26.63	$19.02	40.0%
Bachelor's degree only	$35.81	$28.27	26.7%
Graduate degree	$40.69	$36.22	12.3%

Source: Economics and Statistics Administration calculations using Current Population Survey public-use microdata.

Note: Full-time private wage and salary workers.

STEM Skills in Demand throughout the Economy

Just as innovative processes take place both inside and outside the traditional spheres of research and development (R&D), STEM is now often defined both inside and outside the traditional set of science and engineering jobs. Thus, STEM can be defined not just as a group of workers in science and engineering jobs, but also as a set of workers with STEM education or STEM knowledge and skills, whether or not they work in STEM jobs. The human capital embodied in the work that STEM workers perform is valued in other sectors of the economy. This capital includes knowledge of mathematics, computers, and electronics and more general skills, such as critical thinking, troubling shooting, and various forms of reasoning.[6] More generally, a growing number of occupations in the economy have been found to require a greater intensity of non-routine analytical and interactive tasks—that is, ones requiring reasoning and high executive functioning—while a declining number of occupations rely more heavily on manual and routine tasks.[7]

Nearly two-thirds of workers with undergraduate degrees in a STEM field are working in non-STEM occupations, such as healthcare, education, the social sciences, and management[8] (see figure 4.2). These workers are not underperforming, nor are they mismatched in their current jobs. Rather, the same human capital that drives innovative processes through traditional R&D-related employment is needed across our economy, a suggestion that is confirmed in surveys of these workers as well.[9] Furthermore, many STEM-educated workers who choose education jobs are likely teaching STEM skills to others.

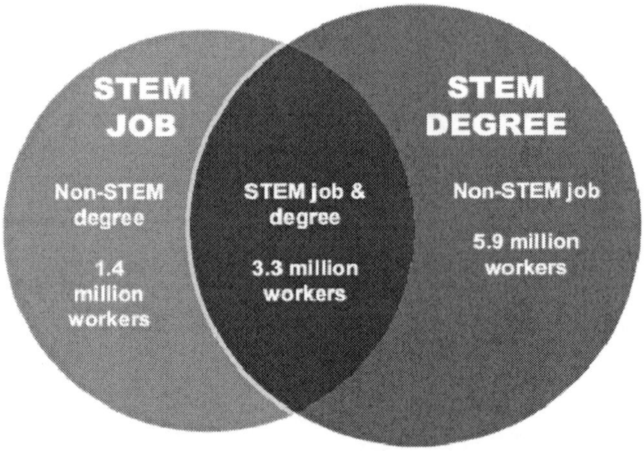

Source: Economics and Statistics Administration calculations from American Community Survey microdata.

Figure 4.2. The Overlap between STEM Jobs and STEM Degrees.

The value of STEM human capital is reflected in the earnings premium enjoyed by college-educated workers with a STEM degree. All else equal, workers with a STEM degree earn 11 percent more per hour in full-time non-STEM jobs than workers with other undergraduate degrees. When STEM majors work in STEM jobs, their earnings premium rises to 20 percent, relative to persons with non-STEM degrees working in non-STEM jobs.[10]

Given that more than two-thirds of STEM workers have at least a college degree and that demand for STEM workers and workers with STEM degrees continues to grow, the U.S. college and university system is a cornerstone of our STEM future. Fortunately, at the college level, the United States continues to set the standard of the quality of the educational system and in the value of obtaining a college degree.

However, the United States is losing ground to other countries in important areas of education, specifically in creating opportunities for students to gain expertise in STEM skills. Improvements are required at all education levels, including post-secondary school, if the United States is to remain internationally competitive and for it to continue to excel in preparing its workforce for an increasingly knowledge-intensive economy.

Many U.S. Universities Are Outstanding but Our Production of U.S. STEM Graduates Is Not

Elite institutions within the United States' college and university system typically dominate global rankings of prestigious higher education institutions. In 2011- 2012, in a worldwide ranking, 18 out of the top 25 universities and 30 out of the top 50 universities were in the United States.

The United Kingdom was next with four in the top 25 and five in the top 50.[11]These rankings make our country a magnet for the best students from around the world. The United States is still the top destination for students studying abroad, although its share has fallen somewhat over time (see figure 4.3).

Another way to look at the desirability of the United States as a destination for study is in export terms: when students from abroad come to the United States to study, that is an export of educational services (see figure 4.4). In 2010, receipts from education exports exceeded $21 billion, more than doubling over the previous 10 years in keeping with the rising cost of attending U.S. colleges and univer sities.

Close to half of the receipts came from China ($4.0 billion), India ($3.3 billion), and Korea ($2.2 billion) (see figure 4.5). Roughly 40 percent of interna tional students in 2010–2011 were studying in STEMrelated fields, such as engi neering (18.7 percent), math and computer sciences (8.9 percent), and physical and life sciences (8.8 percent). Business and management ranked the most popu lar individual field (21.5 percent).[12]

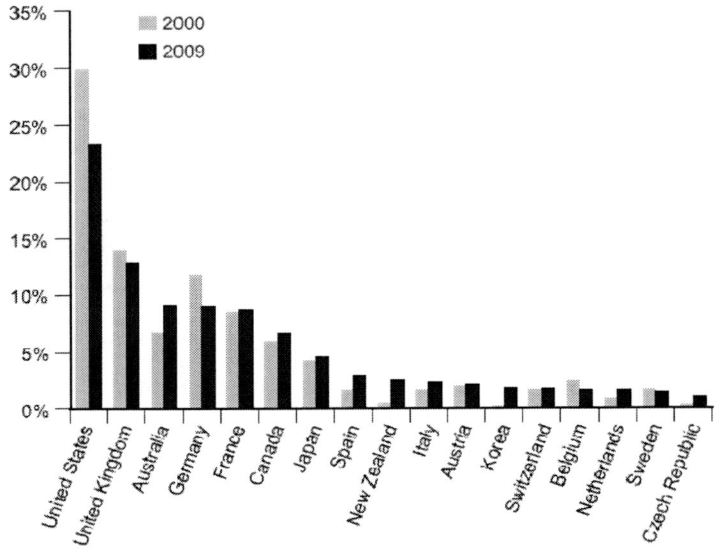

Source: OECD Education at a Glance 2011, Table C3.6.
Note: OECD member countries with fewer than 1% of foreign students are not shown.

Figure 4.3. Distribution of Foreign Tertiary Students Across OECD Countries.

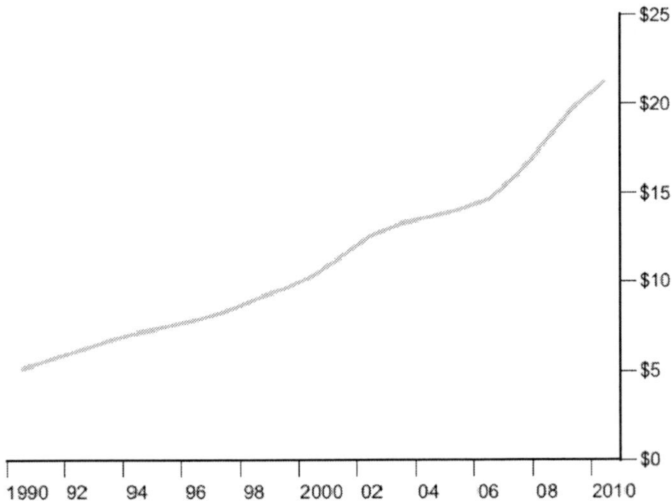

Source: Bureau of Economic Analysis.
Note: Data are in billions of current dollars.

Figure 4.4. Exports of Educational Services, 1990–2010.

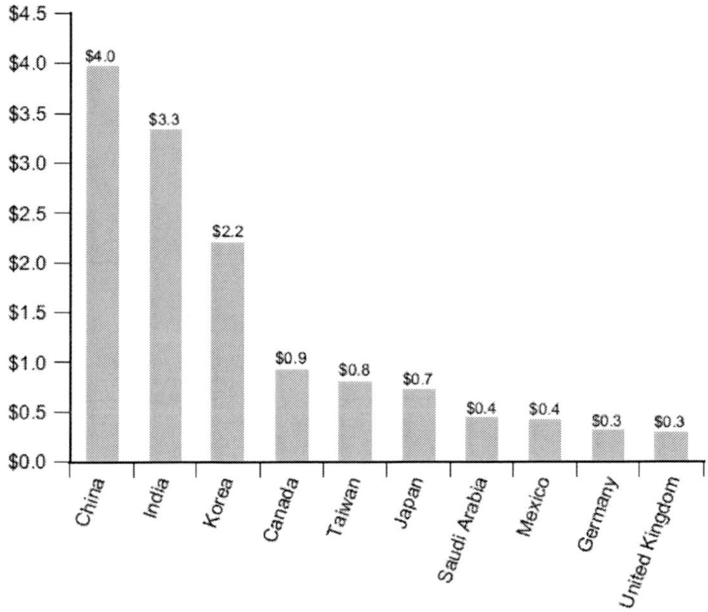

Source: Bureau of Economic Analysis.
Note: Data are for 2010 and are in billions of current dollars.

Figure 4.5. Exports of Educational Services to Selected Countries, 2010.

While the United States continues to have topflight higher education institutions, fundamental problems in the kindergarten through college system threaten our ability to increase the skills of our workforce as rapidly as needed. Among high school graduates who do enroll in college, a remarkably high propor tion—20 percent—takes at least one remedial course their freshman year.[13] Stu dents who take remedial coursework often do not fully catch up with their other collegegoing peers: compared with college students who need no remediation, students who take even a single remedial course are less likely to earn their bach elor's degree than students who did not take any remedial courses.[14] More gen erally, the United States has slipped behind other countries in terms of college attainment rates over the second half of the 20th century. The cohort born be tween 1943 and 1952 had the highest share of bachelor degree holders in the world. Since then several other countries have not only caught up but surpassed the United States in the proportion of adults who have completed college. Currently, the share of the U.S. population aged 25–34 that has attained post-secondary education is only slightly above the OECD average.[15]

Of those who graduate from college, the United States produces fewer STEM graduates relative to other developed countries. OECD data show that in 2009 12.8 percent of U.S. graduates with bachelor's degrees were in STEM fields. This places the United States near the bottom of OECD countries in terms of the percentage of STEM graduates produced. Significant economic competitors—such as South Korea (26.3 percent), Germany (24.5 percent), Canada (19.2 percent), and the United Kingdom (18.1 percent)—are on the long list of countries producing a much higher percentage of STEM graduates.[16]

As they advance through the education system, U.S. students choose not to enter STEM fields or, if they do pursue these studies, do not continue. Three out of four high school students who test in the top math quartile don't start with a STEM major in college, and only half of all students who start in a STEM major graduate with a STEM degree.[17] While no single reason can account for the low share of students in STEM fields, students' poor K–12 math and science preparation and their unwillingness to commit the additional study time needed for math and science courses relative to other classes are likely contributing factors.[18] As detailed below, the Department of Education and the National Science Foundation have developed initiatives to improve K–12 and college-level STEM instruction and to reduce the number of students exiting STEM majors for other majors.

The High Cost of College and Poor Academic Preparation Deter Students

Given the importance of a college education to a worker's productivity and earnings, particularly for STEM-educated workers, it is striking that only 70 percent of high school graduates in 2009 went on to some higher education—a rate lower than that of the highest performing countries, such as Norway and New Zealand.[19]

One barrier to college attendance is the high price of tuition and fees. Whether for a 2-year or 4-year degree, tuition has climbed much faster than consumer prices and household incomes. Over the past decade, in-state public university tuition and fees more than doubled while tuition and fees for 2-year schools rose 71 percent. During the same period, overall consumer prices increased 27 percent and nominal median household income rose 18 percent (see figure 4.6). In other words, household income over the period was not able to keep up with the overall increase in consumer prices, let alone the soaring sticker price of a college education. The cost of room and board (not included in tuition and fees) was no more forgiving. Between the 1999–2000 and 2009–2010 school years, the cost of staying in a college dormitory rose 80 percent while board increased 55 percent. Grant aid from public and private sources, including Federal Pell Grants and Federal education tax credits and deductions, however, have helped soften the financial blow to families. As a result, the net price of a college education—that is, the published price of tuition and fees minus all forms of financial aid—has not increased as fast as the sticker prices.[20] In fact, in constant dollars the net price for full-time students attending public, four-year institutions in 2011–2012 increased just $60 relative to 2007–2008, while the net price for public, two-year schools and private schools in 2011–2012 was lower than in 2007–2008.[21]

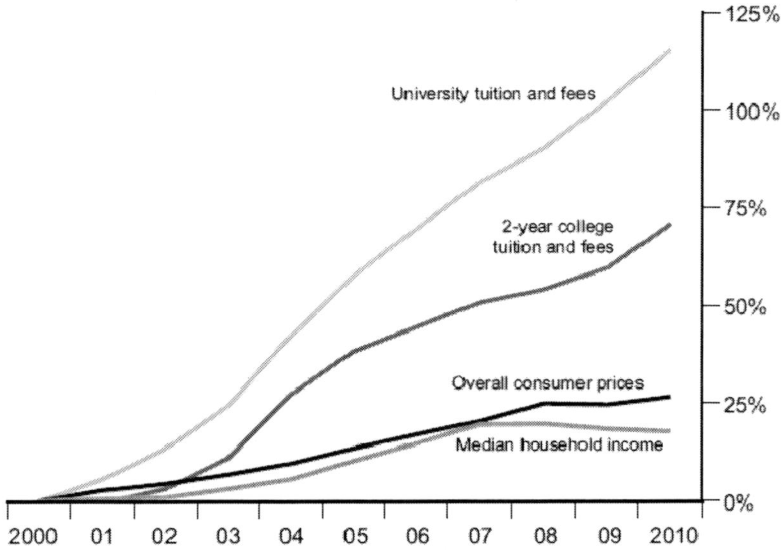

Source: National Center for Education Statistics, Census Bureau, and Bureau of Labor Statistic.

Figure 4.6. Percentage Growth Since 2000 in College Tuition, Consumer Prices, and Median Household Income.

Another barrier to attending college (and an explanation for the high rate of remedial education in college) is inadequate K–12 preparation. The primary and secondary education system in the United States must prepare students who wish to go to college and specialize in a STEM field with the skills to do so. Similarly, those students who choose to enter the workforce directly after high school and not attend college must be equipped with the skills necessary to be trained for STEM jobs that do not require a college education. Yet pre-college preparation in the skills that will allow students to specialize in STEM coursework in college or to enter STEM jobs right out of high school is lagging. The Program for International Student Assessment (PISA) test scores reveal that U.S. students consistently scored below the OECD average in math in 2003, 2006, and 2009 (the past three testing cycles). In science, while U.S. students scored lower than the OECD average in science literacy in 2006, the average score of U.S. students in 2009 was not measurably different from the 2009 OECD average (see figure 4.7). Further, U.S. students scored lower than the students in 12 OECD countries, and not significantly different from students in 12 other countries. These conclusions about the world ranking of U.S. students is supported by the results of the most recent National Assessment for Educational Progress study, which shows that although U.S. students have improved in math over the past 30 years, only 26 percent of 12th graders are "proficient" or better in math. In reading, 38 percent of students scored at the proficient level or higher in 2009. While overall math and reading scores for 12th graders have improved between 2005 and 2009 (the latest two reports available), there remain notable and persistent disparities by race, ethnicity and gender.[22]The latest science scores may also give reason for pause as only 21 percent of 12th graders were found to be "proficient" or better.[23] Overall these scores suggest that while we need to boost student achievement in all dimensions, we are particularly poor right now in skills that prepare students for post-secondary STEM education and training.

Although post-secondary education is the principal path into a STEM job, a 4-year degree is just one option for future or current workers who want to gain STEM-related knowledge and skills. With relatively low tuition, wide dispersion through the United States, convenient class times, and course offerings aimed at students from diverse high school backgrounds, our nation's community colleges lower the barriers to post-secondary education. A recent study of Florida community colleges highlights their dual role in increasing economic mobility by enabling students (particularly low-income students with good grades in high school) to transfer to 4-year colleges and in teaching work-enhancing skills (which particularly benefit low-income students who were less successful in high school).[24] As the Florida study and others highlight, the payoff of choosing more technically oriented fields is considerable. This becomes particularly clear when examining training programs aimed at dislocated workers, for whom 1 year of technical training can increase workers' re-employment earnings by $1,600, compared with $800 for other types of training.[25] Note that these results related to just 1 year of study, as opposed to a 2-year degree.

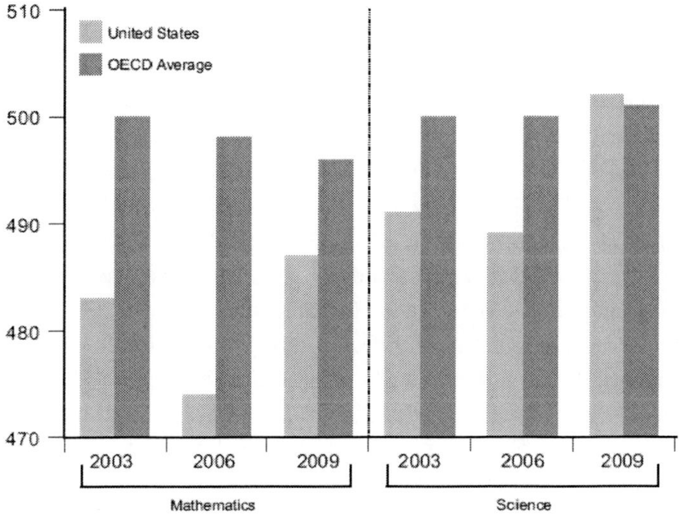

Source: OECD, PISA 2003, 2006, and 2009 databases.

Figure 4.7. Math and Science Test Scores in the U.S. and OECD.

Demographics Create Challenges and Opportunities for Growth

Given the advantages of working in a STEM occupation and having an educational background in STEM, there are disturbing demographic disparities in STEM education and in the composition of workers in STEM occupations. Women are vastly underrepresented among STEM workers. Despite making up nearly half of the U.S. workforce, women hold less than 25 percent of STEM jobs, and this disparity has persisted throughout the past decade, even as college-educated women have increased their share of the overall workforce (see figure 4.8). Though this varies by field of study, overall women hold a disproportionately low share of STEM undergraduate degrees. For example, this is particularly true in engineering, though women receive the majority of degrees in biology. Also, women with a STEM degree are less likely than their male counterparts to work in a STEM occupation and more likely to work in education or healthcare. This has real consequences, as women with STEM jobs earned 33 percent more than comparable women in non-STEM jobs—considerably higher than the STEM premium for men—so the gender wage gap is smaller in STEM jobs than in non-STEM jobs.[26]

Like women, most racial and ethnic minorities are underrepresented among STEM workers. A noticeable exception is non-Hispanic Asians. Fifteen percent of all non-Hispanic Asians work in STEM jobs, almost 3 times the overall share of STEM workers in the economy. This reflects non-Hispanic Asian's greater likelihood of graduating from college, majoring in a STEM discipline, and working in a STEM job given a degree in a STEM major. For example, non-Hispanic Asians are most likely (42 percent) to graduate from college with a STEM degree, while the propensities of other groups are all fairly similar (17–22 percent). Half of all non-Hispanic Asian workers with STEM degrees have STEM jobs, compared to 30 percent of Hispanics, non-Hispanic black, and American Indian and Alaska Native workers. Interestingly, on average, all minority groups have higher wage premiums from having a STEM job than do non-Hispanic whites (31 to 39 percent versus 22 percent). With greater equality in educational attainment, demographic disparities within the STEM workforce can be diminished, helping to boost STEM employment and U.S. leadership in technology and innovation.[2]

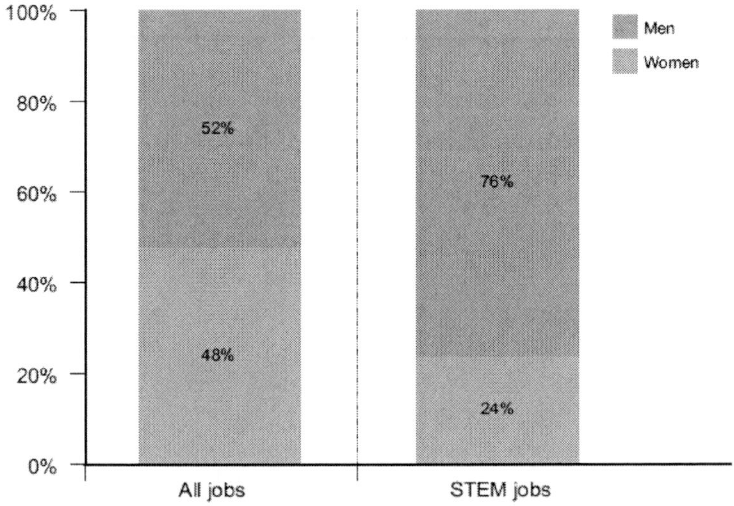

Source: Economics and Statistics Administration calculations from American Community Survey microdata.

Figure 4.8. Gender Distribution Between STEM and All Employment, 2009.

The Foreign-Born Are Key Members of the STEM Workforce

Many innovations that were born in America have been developed by persons who were not born in America. One in five STEM workers is foreign born, with 63 percent coming from Asia. The foreign-born share of STEM workers with graduate degrees (44 percent) is about twice the foreign-born share of STEM workers for all education levels and has nearly doubled over the past 17 years, as has the foreign-born share of STEM workers with just a bachelor's degree also has posted strong gains[28] (see figure 4.9).

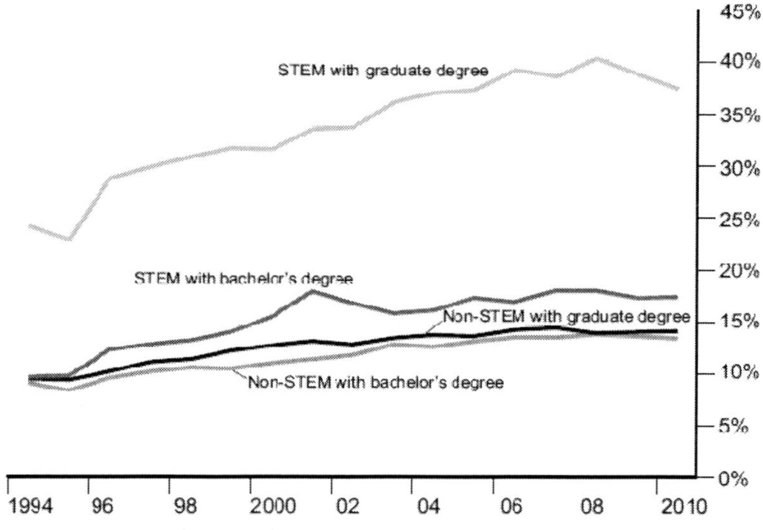

Source: Economics and Statistics Administration calculations using Current Population Survey public-use microdata.

Figure 4.9. Foreign-Born Share of STEM and Non-STEM Employment, by Education.

The growth in the foreign-born STEM workforce reflects multiple factors affecting the supply of and demand for STEM workers. One factor is the difficulty that employers often report in finding applicants with the right technical skills to fill their job openings. Even as we emerge from a historically deep recession, employers report shortages of skilled workers including engineers and software developers.[29]

In a global economy, the payoff to attracting the brightest minds to the United States has been considerable. Consider, for example, that nearly 20 percent of the *Fortune 500* firms founded between 1985 and 2010 were started by an immigrant to the United States.[30]

Many of the foreign-born students educated in STEM disciplines in the United States want to remain here lawfully—starting their own firms or contributing to the growth of existing firms. The United States must develop immigration policies to ensure that this country is welcoming to the world's best and brightest.

The Administration Is Lowering the Barriers to a College Education

States and localities, like American families, face difficult budget situations following the recent deep recession. This has led to difficult choices regarding education. The Obama Administration recognizes these difficulties and has worked on several fronts to make critical investments in our education system—investments that make college affordable and increase the quality and payoff of the education investment that American families are making. These initiatives will strengthen our future and current workforce and more fundamentally build our overall innovative capacity.

Making College More Affordable

Since its origin in 1972, the Federal Pell Grant program has become the most significant source of Federal grant aid to college students and the largest single source of aid at public colleges and universities. The Obama Administration has worked to raise both the maximum Pell Grant amount and expand the number of grants awarded. Through amendments to Higher Education Act of 1965 (HEA) by the American Recovery and Reinvestment Act (ARRA) and the Student Aid and Fiscal Responsibility Act (SAFRA), the maximum Pell Grant award was raised from $4,731 in 2008 to $5,550 in 2010. Beginning in 2013, the maximum Pell grant will increase with the Consumer Price Index. SAFRA also made Federal loans available directly to students, ending wasteful subsidies once paid to lenders and other state guaranty agencies. Overall, the Administration has maintained extraordinary commitment to the Pell program, with total aid to students increasing from $18 billion in 2008 to more than $30 billion in 2011.[31]

These initiatives have succeeded in holding down the growth in the out-of-pocket costs students and their families are paying for college. Over 9 million college students received an average of $3,700 in Pell Grant awards in the most recent academic year, as compared to 5.5 million college students who received an average Pell Grant award of $2,650 in the year before President Obama took office.[32]

In addition to expanding and increasing Pell Grant availability and awards, ARRA established the American Opportunity Tax Credit (AOTC), which provides up to $2,500 a year for college tuition and related expenses for American families. This tax credit improves notably upon the Hope Scholarship credit that it replaced. AOTC has a higher maximum benefit, and it can be claimed for up to four years rather than only two years of undergraduate education. Furthermore, AOTC has a higher income eligibility cutoff, thus making it available to more middle-class families, and it is partially refundable, making it more beneficial to lower-income families. This credit was expected to benefit 9.4 million students and their families in 2011. In December 2010, the President signed an extension of the AOTC through the end of 2012.

The SAFRA Act also greatly improved the terms of an income-based repayment program established in 2007 for student loans. Under these improvements, borrowers will have their student loan payments capped at 10 percent rather than 15 percent of their discretionary income. This new cap was originally going to be available only to new borrowers after July 1, 2014, but President Obama recently announced the availability of a similar "pay-as-you-earn" plan two years earlier. Borrowers who keep up their payments for 20 years will see their remaining debts forgiven—or 10 years for persons with public service jobs.[33]

Addressing STEM Shortcomings

To address the poor STEM participation and performance in our nation's schools, the Administration has launched multiple initiatives (see box 4.1 for a discussion on the efforts mandated by COMPETES to develop an inventory of all STEM educational initiatives). "Educate to Innovate" establishes five major public-private partnerships to harness the power of media, interactive games, hands-on learning, and community volunteers to reach millions of students and expand STEM education and opportunities to all students, particularly those of underrepresented groups.

A necessary step to improving our students' understanding of STEM fields, which should, in turn, lead to more college graduates with STEM training and more STEM workers, is to train additional STEM teachers. Of course, having more teachers is only effective if it does, in fact, lead to an increase in college graduation rates in STEM fields. The Widening

Implementation and Demonstration of Evidence based Reforms (WIDER) program at NSF will help improve undergraduate STEM instruction and outcomes at universities.

Finally, STEM education and career opportunities for underrepresented groups, including minorities and women and girls, need to be expanded. To this end, the "NSF Career-Life Balance Initiative," has been announced. This is a 10-year plan designed to give flexibility to women and men who pursue research careers. For example, NSF will expand a program that will allow researchers to delay or suspend their grants for up to one year in order to care for a newborn or newly adopted child or fulfill other family obligations.

Box 4.1. Inventory of Federal STEM Educational Programs

Section 101 of COMPETES requires the White House Office of Science and Technology Policy (OSTP) to prepare an annual report to Congress describing STEM educational programs and activities by Federal agency in the prior and current fiscal years as well as in the President's budget.[1] The report will also list the programs' funding levels, evaluate their duplication and fragmentation, and describe how participating Federal agencies will disseminate information about federally supported resources to STEM educators. In partial fulfillment of this requirement, OSTP has developed a detailed inventory covering all 13 Federal agencies that sponsor such programs. The inventory tallied 252 specific programs with a total Federal investment of $3.5 billion. About $1 billion of that is being spent to train individuals for activities specific to the mission of those funding agencies, including National Institutes of Health training programs to help develop the next generation of biomedical researchers and US Department of Agriculture programs to train agricultural scientists.

[1] Office of Science and Technology Policy Press Release, "Federal Science, Technology, Engineering, and Math Education Inventory Highlighted," September 19, 2011; www.whitehouse.gov/sites/default/files/microsites/ostp/ostp-stem-inventory_9-19-11.pdf, and America COMPETES Reauthorization Act of 2010, Pub L. No. 115–358, January 4, 2011; www.gpo.gov/fdsys/pkg/PLAW-111publ358/pdf/PLAW-111publ358.pdf.

Helping Community Colleges Assist Workers and Businesses

The Health Care and Education Reconciliation Act (HCRA) includes a $2 billion investment in our nation's community colleges, enabling eligible institutions of higher education to expand their capacity to provide quality education and training services to Trade Adjustment Assistance (TAA) eligible workers as well as other individuals to improve their knowledge and skills and enable them to obtain high-quality employment. Already $500 million in grants have been awarded to community colleges around the country to expand and improve their ability to deliver education and career training programs that can be completed in two years or less. These grants support partnerships between community colleges, community organizations, and employers to develop programs that provide pathways to good jobs, including building instructional programs that meet specific industry needs.

Further serving displaced workers, the Skills for America's Future initiative, an industry-led initiative announced in October 2010, will build and improve partnerships between businesses and educational institutions to train American workers for 21st century jobs. The initiative was created to foster collaborative efforts between the private sector, community colleges, labor unions, and other institutions, with a commitment to scaling up meaningful and measurable solutions. The goal is to build a nationwide network of stakeholders who will work to maximize workforce development strategies, job training programs, and job placement. The Skills for America's Future Task Force has been created and co-chaired by top-level Administration policymakers, to coordinate Federal efforts.[34]

The Race to the Top Initiative Rewards Statewide Reform

The Race to the Top Fund uses competitive grants to encourage comprehensive state and local reform that result in increased student achievement, narrowed achievement gaps, and improved high school graduation and college enrollment rates.[35] As part of Race to the Top, the Department of Education has awarded $4 billion in competitive grants to 11 states and the District of Columbia over two phases that will directly impact 13.6 million students and 980,000 teachers in 25,000 schools.[36] An additional $700 million was made available in 2011, $200 million of which was used to make additional awards to enable states to carry out meaningful portions of their ambitious reform plans. The remaining $500 million was awarded to nine states through for the new Race to the Top Early Learning Challenge, a competition to support the states with the most ambitious plans to ensure that high-need children from birth to age five enter kindergarten ready to succeed.[37]

Enhancing Our Nation's Educational Infrastructure

As the United States emerges from the Great Recession, states and localities still face reduced revenues and are continuing to reduce budgets. Local schools, for example, cut nearly 235,000 jobs from May 2009 to November 2011. At the same time, budgets to maintain our nation's more than 100,000 public schools have been pared back, which has led to a $270 billion backlog of deferred maintenance and repair. The cost of heating and cooling antiquated and inefficient buildings lead districts to spend more each year on their energy bills than on computers and textbooks combined. Increasing class sizes combined with aging buildings result in overcrowded schools that have crumbling ceilings and inadequate wiring to support today's information technology infrastructure. More funds are needed to enhance our public schools, with a priority placed on high-need and rural schools, Bureau of Indian Education schools and community colleges (including tribal colleges).

References

Adelman, Clifford. 1998. "The Kiss of Death? An Alternative View of College Remediation." *National Crosstalk*; www.highereducation.org/crosstalk/ct0798/voices0798-adelman.shtml.

Baumol, William J., Melissa A. Schilling, and Edward N. Wolff. 2009. "The Superstar Inventors and Entrepreneurs: How Were They Educated?" *Journal of Economics and Management Strategy* 18, no. 3 (Fall): 711–728.

Becker, Gary S. 2008. "Human Capital." *The Concise Encyclopedia of Economics*. Library of Economics and Liberty. Accessed December 4, 2011; www.econlib.org/library/Enc/HumanCapital.html.

Beede, David, Tiffany Julian, Beethika Khan, Rebecca Lehrman, George McKittrick, David Langdon, and Mark Doms. 2011a. *Education Supports Racial and Ethnic Equality in STEM.* ESA Issue Brief 05–11. Washington, DC: U.S. Department of Commerce, Economics and Statistics Administration (ESA), September; www.esa.doc.gov/sites/default/files/reports/documents/educationsupportsracialandethnicequalityinstem_0.pdf.

Beede, David, Tiffany Julian, David Langdon, George McKittrick, Beethika Khan, and Mark Doms. 2011b. *Women in STEM: A Gender Gap to Innovation.* ESA Issue Brief 04–11. Washington, DC: U.S. Department of

Commerce, ESA, August; www.esa.doc.gov/sites/default/files/reports /documents/womeninstemagaptoinnovation8311.pdf.

Carnevale, Anthony, Nicole Smith, and Michelle Melton. 2011. *STEM.* Report. Washington, DC: Center on Education and the Workforce, Georgetown University, October; www9.georgetown.edu/grad/gppi/hpi /cew/pdfs/stem -complete.pdf.

College Board. 2011. *Trends in College Pricing 2011.* New York, NY: Advocacy and Policy Center; trends.college board.org/college_pricing /indicator/index.

Council of Economic Advisers. 2009. *Preparing the Workers of Today for the Jobs of Tomorrow.* July; www.white house.gov/administration/eop/cea /Jobs-of-the-Future.

Council of Economic Advisers. 2010. *Economic Report of the President.* Washington, DC: U.S. Government Printing Office, February; www.white house.gov/sites/default/files/microsites/economic-report-president.pdf.

Council of Economic Advisers. 2011a. *The Economic Report of the President.* Washington, DC: U.S. Government Printing Office, February; www.gpoaccess.gov/eop/2011/pdf/ERP-2011.pdf.

Council of Economic Advisers. 2011b. *Making College More Affordable: Implications of New Data.* October 26; www.whitehouse.gov/sites/default /files/20111026-cea-report-making-college-more-affordable.pdf.

Federal Reserve Board. 2011. "Summary." In *The Beige Book.* Report for September 7, 2011, www.federalreserve.gov/fomc/beigebook/2011 /20110907/default.htm and Report for October 19, 2011, www.federal reserve.gov /fomc/beigebook/2011/20111019/default.htm.

Furchtgott-Roth, Diana, Louis Jacobson, and Christine Mokher. 2009. *Strengthening Community Colleges' Influence on Economic Mobility.* Economic Mobility Project, Pew Charitable Trusts, October; www.economicmobility. org/assets/pdfs/PEW_EMP_COMMUNITY_COLLEGES.pdf.

Goldin, Claudia, and Lawrence F. Katz. 2008. *The Race between Education and Technology.* Cambridge, MA: Harvard University Press.

The Hamilton Project. 2010. *Retraining Displaced Workers.* Policy Brief. Washington, DC: The Brookings Institution, October; www.brookings.edu /~/media/Files/rc/papers/2010/10_renew_communities_greenstone_looney / 10_displaced_workers_lalonde.pdf.

Institute of International Education. 2011. *Open Doors Data*, 2009–2010. Accessed November 28, 2011; www.iie.org/Research-and-Publications /Open-Doors/Data/International-Students/Field-of-Study/2008-10.aspx.

Partnership for a New American Economy. 2011. "The 'New American' Fortune 500." June 2011; www.renewour economy.org/sites/all/themes/pnae/img/new-american-fortune-500-june-2011.pdf.

Langdon, David , George McKittrick, David Beede, Beethika Khan, and Mark Doms. 2011. *STEM: Good Jobs Now and for the Future.* ESA Issue Brief 03–11. Washington, DC: U.S. Department of Commerce, ESA; www.esa.doc.gov/sites/default/files/reports/documents/stemfinalyjuly14_1.pdf.

National Center for Education Statistics. 2010a. "Chapter 2: Postsecondary Education." *2010 Digest of Education Statistics;* nces.ed.gov/programs/digest/d10/ch_3.asp.

National Center for Education Statistics. 2010b. "The Nation's Report Card: Grade 12 Reading and Mathematics 2009 National and Pilot State Results." *National Assessment of Educational Progress.* November 2010; nces.ed. gov/nationsreportcard/pdf/main2009/2011455.pdf.

National Center for Education Statistics. 2011. "The Nation's Report Card: Science 2009." *National Assessment of Educational Progress at Grades 4, 8, and 12.* January 2011; nces.ed.gov/nationsreportcard/pdf/main2009/2011451.pdf.

National Science Foundation. 2010. "Science and Engineering Labor Force." In *Science and Engineering Indicators: 2010*; www.nsf.gov/statistics/seind10/pdfstart.htm.

Organisation of Economic Co-operation and Development (OECD). 2011a. *Education at a Glance 2011: OECD Indicators*; www.oecd.org/dataoecd/61/2/48631582.pdf.

Organisation of Economic Co-operation and Development (OECD). 2011b. "Graduates by Field of Education." Accessed November 28, 2011 at *OECD.StatExtract;* stats.oecd.org/Index.aspx?DatasetCode=RGRADSTY.

Silverman, Rachel Emma, and Joe Light. 2011. "Generation Jobless: Students Pick Easier Majors Despite Less Pay." *Wall Street Journal,* November 9; online.wsj.com/article/SB10001424052970203733504577026212798573518.html.

Times Higher Education. 2011. *World University Rankings 2011–2012.* Thomson Reuters. Accessed December 28; www.timeshighereducation.co.uk/world-university-rankings.

U.S. Department of Education. *Race to the Top Fund.* Accessed November 28, 2011; www2.ed.gov/programs/ racetothetop/index.html.

U.S. Department of Education. 2011a. "U.S. Department of Education: The FY 2010 Summary of Performance and Financial Information." February 2011; www2.ed.gov/about/reports/annual/2010report/summary.pdf.

U.S. Department of Education. 2011b. "2009–2010 Federal Pell Grant Program End-of-Year Report." *Office of Postsecondary Education.* May 2011; www2.ed.gov/finaid/prof/resources/data/pell-2009-10/pell-eoy-09-10.pdf.

U.S. Department of Education. 2011c. "ED Review." May 2011; www2.ed.gov/news/newsletters/edreview/2011/ 0527.html.

White House. 2010. "Ensuring that Student Loans are Affordable." Fact Sheet. January; www.whitehouse.gov/sites/default/files/100326-ibr-fact-sheet.pdf.

INFRASTRUCTURE FOR THE 21ST CENTURY

> "It's the broadband networks beneath us and the wireless signals around us, the local networks in our schools and hospitals and businesses, and the massive grids that power our nation. It's the classified military and intelligence networks that keep us safe, and the World Wide Web that has made us more interconnected than at any time in human history."
>
> *—President Barack Obama, Remarks on Securing Our Nation's Cyber Infrastructure, May 29, 2009*

Introduction

Providing and maintaining infrastructure is one of government's most important roles, and perhaps one of the most underappreciated. Infrastructure improves the lives of individuals every day, providing electricity and water, the roads and public transportation needed for commuting and shopping, and the telecommunications networks needed for the free flow of information and ideas. Businesses rely on that same infrastructure to interact with suppliers, to produce their goods and services, and to provide those goods and services to their customers. In short, "infrastructure—freight ports, airports, bridges, roads, rail and transit networks, water and sewer systems, web of channel communications—is the connective tissue of our nation" and it "has a dramatic effect on the economic competitiveness of our nation, the health of our environment and our quality of life."[1]

In the past, the United States led the way in several key areas of infrastructure development, starting with the railroad system of the 1800s, an innovation that greatly reduced travel times and allowed more robust commerce between the states and the rest of the world. During the 20th century, electricity was brought to nearly all homes and businesses; the telephone system brought the ability to communicate instantly to virtually everyone; paved roads and highways were vastly increased, with the Eisenhower's Interstate Highway System at the center; and an air transportation system emerged to enable unprecedented personal mobility and access to global products and services. The United States must still be mindful about the condition of its existing infrastructure, as these investments remain crucial to its economic health and investments in these assets can provide high levels of return. Thus, repairing roads and bridges and maintaining the energy grid and the telecommunications networks need to remain priorities. Furthermore, these crucial investments in repairing and rebuilding the existing infrastructure should incorporate principles of sustainable design.

However, in today's economy—with global supply chains, exports to foreign markets, telework, and just-in-time inventories—the nature of infrastructure needed to compete is changing, and the United States needs a 21st century infrastructure to ensure that it remains competitive. This includes improvements to existing infrastructure, such as introducing advances into the highway system that will provide safety, mobility and energy efficiency. Infrastructure for the 21st century is led by the ever-growing presence and influence of broadband Internet. Definitions of what constitutes "broadband" can differ, but one useful measure put forth by the Federal Communications Commission (FCC) in its recent National Broadband Plan sets a goal of ensuring that 100 million households have access to broadband Internet with actual download speeds of 50 megabits per second (mbps) and actual upload speeds of 20 mbps by 2015, and that 100 million households should have access to actual download speeds of 100 mbps and actual upload speeds of 50 mbps by 2020. In addition, the FCC has set as a goal that every community should have affordable access to service with speeds of at least one gigabit per second at schools, hospitals and government buildings.[2]

Broadband Internet is used by more than 2 billion people and is still growing. Internet-related consumption and expenditure is now bigger than agriculture or energy. Research has shown that the Internet accounts for, on average, 3.4 percent of GDP across many major countries, reaching as much as 6 percent of GDP in advanced countries such as Sweden and the United Kingdom.[3] The Internet is also a critical element of growth; in a study of mature economies, the Internet accounted for 10 percent of GDP growth over the past 15 years, and this contribution has been increasing over time; the Internet's contribution to GDP growth was 21 percent in the last five years.[4]

Small and medium-sized enterprises (SMEs), in particular, have benefitted from the Internet. SMEs with a strong web presence have been found to grow faster and export more than those that had minimal or no presence. One survey found these firms also created more than twice the number of jobs as firms without an Internet presence, creating 2.6 jobs for each one eliminated.[5] The Internet also has created large amounts of consumer value. For example, it is estimated that the Internet generated $64 billion in consumer value in the United States in 2009.[6]

Also, a strong 21st century infrastructure is crucial because it is closely linked to the other key building blocks of competitiveness. Additional basic and applied research will lead to improvements in information infrastructure, and that infrastructure is necessary to ensure that advances in research of all types can be efficiently shared among the research community. Education is also closely linked to infrastructure. Not only does traditional infrastructure, namely school buildings, need to be in good shape in order for students to learn, but information infrastructure, such as broadband, can help deliver new education technologies to the classroom. Although private industry has led the way in many forms of infrastructure (most utilities are privately owned companies, for example), government maintains an important role because infrastructure can have positive spillover effects that a private investor would not take into account when deciding whether or not to invest in the project. As with R&D, it is likely that too little money would be spent on infrastructure without government intervention.[7] In addition, our society has affirmed repeatedly that we would like all of our citizens to have access to certain technologies. In the past, this meant providing phone access or electricity to all parts of the country. Today, expanded high-speed Internet access would not be available to certain areas without government assistance. As discussed below, though the United States has made great strides in harnessing the transformational and economic power of the Internet and other aspects of a strong 21st century infrastructure, there remains substantial untapped potential.

Definition of 21st Century Infrastructure

Infrastructure is a broad concept and there is no single commonly accepted definition. Traditionally, infrastructure refers to the physical pathways used to transport goods, people, and basic utility services. Most commonly, this includes roads, bridges, seaports, airports, rail lines, the electrical grid, pipelines, and the water and sewage system.

However, more recently, our economy and society have become more information intensive and "information infrastructure" has gained in importance.

Until relatively recently, information infrastructure was limited to voice communication over the physical wires of our telephone system but now includes our broadband Internet connections (including fiber, cable modem and digital subscriber line (DSL) service) as well as satellites and cell phone towers.

Further, information infrastructure also encompasses our burgeoning "cloud computing" capabilities and traditional forms of infrastructure that can be improved when used in conjunction with 21st century infrastructure, such as the "Smart Grid" for electricity.

How Does Our 21st Century Infrastructure Stack up?

Electricity Grid

Much of our nation's traditional electricity infrastructure has changed little from its original design and form at the end of the 19th century.[8] The result, according to U.S. Energy Secretary Steven Chu, is that "the ability of the United States to meet the growing demand for reliable electricity is challenged."[9] Successfully addressing this challenge is critical.

"America cannot," Secretary Chu has stated, "build a 21st century economy with a 20th century electricity system. By working with states, industry leaders, and the private sector, we can build a clean, smart, national electricity system that will create jobs, reduce energy use, and expand renewable energy production."[10]

President Obama has outlined a vision for doubling America's use of clean energy by 2035 and achieving the goal of putting one million electric vehicles on the road by 2015.

Having a modernized, smarter electric system is an important step to meeting these goals.

Building the necessary transmission infrastructure and utilizing smart grid technologies will facilitate the integration of renewable resources into the grid, accommodate a growing number of electric vehicles, help avoid blackouts, restore power more quickly when outages occur, and reduce the need for new power plants. Smart grid technologies also provide a foundation for innovation by entrepreneurs and others who can develop tools to empower consumers and help them make informed decisions about energy usage.

To lay out a path forward, the Federal government, in June 2011, released *A Pol icy Framework for the 21st Century Grid: Enabling Our Secure Energy Future.*[11]

This framework features four pillars, which are supported by Administration actions, and includes further policy recommendations to promote investment, innovation, and job growth:

1) **Enabling cost-effective smart grid investments** by disseminating lessons learned from American Recovery and Reinvestment Act investments.
2) **Unlocking the potential of innovation in the electricity sector** through a greater focus on standards and interoperability (see box 5.1).
3) **Empowering consumers and enabling informed decision making** with enhanced information to save energy, ensure privacy, and shrink bills.
4) **Securing the grid** from cyber attacks and improving its recoverability in the event of such an attack.

The Administration is taking a number of concrete steps to put these principles into action. For example:

- The Department of Agriculture's Rural Utility Service is investing up to $250 million in cost-effective smart grid technology in rural America by June 2012.
- The Administration has also launched Energy.Data.gov, an open government platform that brings together free high-value datasets and tools, maps, and apps to increase awareness of and deepen insights into our nation's energy performance.
- The Department of Defense, through its "microgrid" initiatives, is working to test distributed generation and electricity distribution systems that will help enhance the security and efficiency of military bases both at home and abroad.

Box 5.1. NIST and the Smart Grid[1]

Deploying an interoperable and secure Smart Grid cannot be accomplished without establishing interoperability standards. To accelerate the development of these standards, Congress, under the Energy Independence and Security Act of 2007 (EISA), assigned the National Institute of Standards and Technology (NIST) the "primary responsibility to coordinate development of a framework that includes protocols and model standards for information management to achieve interoperability of Smart Grid devices and systems..." [EISA Title XIII, Section 1305].

Recent Accomplishments

The Smart Grid Program has made significant progress in its mission to coordinate the development of interoperability standards over the course of its history. Below is an abbreviated list of recent accomplishments:

- NIST established the Smart Grid Interoperability Panel in November 2009 as a public-private partnership to provide technical support and provide an open forum for Smart Grid stakeholders.
- NIST published the Framework and Roadmap for Smart Grid Interoperability, an initial list of 75 interoperability standards, in January 2010.
- NIST published Guidelines for Smart Grid Cyber Security in September 2010, providing the technical background and details that inform organizations' efforts to securely implement Smart Grid technologies.
- NIST created a Smart Grid Advisory Committee in September 2010 to provide input on Smart Grid standards and NIST's research activities in this area.
- In July 2011, the Smart Grid Interoperability Panel approved the first set of Smart Grid standards, including standards focused on Internet protocols, energy usage information, vehicle charging stations, smart meter upgradeability, and wireless communication devices.

[1] For more information go to www.nist.gov/smartgrid/index.cfm.

To empower consumers and foster innovation, the Administration will continue to promote the Green Button challenge to industry—the common-sense idea that consumers should be able to get access to and download their own energy usage information. And finally, the Administration is working to improve the overall quality and timeliness of electric transmission infrastructure permitting through the interagency Rapid Response Team for Transmission.

Information Infrastructure

Recent developments in information and communications technology (ICT), such as high-speed or broadband Internet access have transformed the social and economic environment in which we live. It provides an outstanding channel for gathering and disseminating information, entertainment, commerce, and education, and it can bring substantial benefits to our economy. For example, "electronic health records could alone save more than $500 billion over 15 years.

Much of the electric grid is not connected to broadband, even though a Smart Grid could prevent 360 million metric tons of carbon emissions per year by 2030, equivalent to taking 65 million of today's cars off the road. Online courses can dramatically reduce the time required to learn a subject while greatly increasing course completion rates."[12]

In the United States, 68 percent of households had broadband Internet access in 2010, an almost eight-fold increase since 2001.[13] However, there are still a large number of Americans who do not use or lack access to the latest broadband and information technologies; a 68 percent adoption rate still leaves approximately one-third of American homes cut off from the digital economy. Furthermore, despite impressive gains, the United States trails behind a number of other nations in household-level broadband adoption[14] (see figure 5.1).

Moreover, the use of broadband continues to vary substantially across different demographic groups and geographic areas. People with lower household incomes, people with less education, those with disabilities, as well as Blacks, Hispanics, and rural residents generally lag in broadband use (see table 5.1).

Bridging these socioeconomic gaps in household-level broadband connectivity in the United States is critical to capturing the potential economic and social gains that accompany more widespread adoption of broadband services.

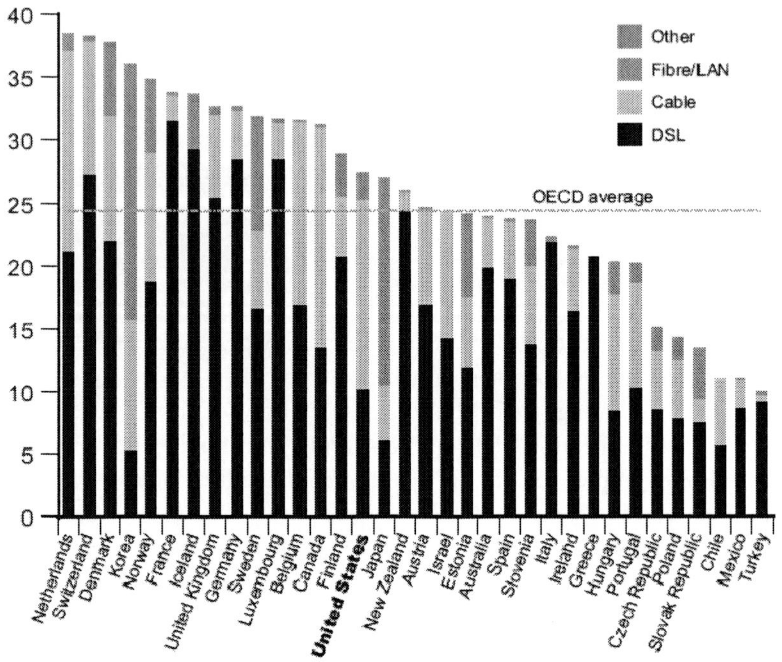

Source: OECD Broadband Portal, Table 1d, June 2011.

Figure 5.1. OECD Fixed (Wired) Broadband Subscriptions per 100 Inhabitants.

At least part of these disparities is likely explained by the fact that, in some cases, particularly in rural parts of the country, broadband is simply not available,[15] largely due to the economic challenges faced by commercial providers seeking to profitably serve large geographic areas that are sparsely populated.

The government has sought to address these needs, typically with grant and loan programs to providers such as the Broadband Technology Opportunities Program (BTOP) at the Department of Commerce, which has been successful in extending broadband to under-served communities (see box 5.2). Further, the FCC recently revised its universal service fund, which traditionally has subsidized the provision of basic telephone service to rural and high-cost areas, so that it will begin to support broadband service by commercial providers in those areas.[16] The lessons from these policy efforts should prove instructive as public and private stakeholders continue to bring broadband to more Americans.

Table 5.1. Household Computer and Broadband Use

	Computer Use		Broadband Access	
	Urban	**Rural**	**Urban**	**Rural**
All households	78%	70%	70%	57%
	Race and Ethnicity			
White, non-Hispanic	82%	72%	75%	60%
Black, non-Hispanic	66%	53%	57%	41%
Hispanic	67%	57%	58%	46%
	Household Income			
$25,000 or less	56%	49%	45%	35%
$25,001–$50,000	76%	74%	67%	60%
$50,001–$75,000	88%	87%	82%	76%
$75,001–$100,000	93%	91%	89%	82%
$100,001 or more	96%	94%	93%	87%
	Education			
Less than a high school degree	46%	39%	35%	26%
High school degree	68%	64%	59%	50%
Some college	84%	82%	75%	69%
College degree or more	93%	89%	88%	80%

Source: Economics and Statistics Administration and National Telecommunication and Information Administra tion. 2011. *Exploring the Digital Nation: Computer and Internet Use at Home.* Washington, D.C: U.S. Department of Commerce, November 2011; www.esa.doc.gov/sites/default/files/reports/documents /exploringthedigitalnation-computerandinternetuseathome.pdf.

Broadband also can be provided wirelessly, and the rapid growth of mobile communications clearly shows how important this technology has become to the American way of life. Wireless broadband, like wired broadband, has the potential to transform many different areas of the American economy by providing a platform for new innovation. The spread of wireless broadband will increase the rate of growth in per capita income and will spur economic activity through new business investment. There is the potential for many new high-quality jobs to be created, both directly through investments in wireless infrastructure, and indirectly through as yet unanticipated applications, services and more rapid innovation enabled by advanced wireless platforms.

Although these effects are difficult to quantify precisely, evidence from the economics literature suggests that they are likely to be substantial. Areas where innovations using wireless technologies are likely to have significant effects include consumer products and services; products to enhance business productivity, including business process re-engineering; health care, through products like patient-physician video conferencing, personal handheld biosensors to generate diagnostic information, and remote transmission of diagnostic information and images; education; and public safety, where a nationwide interoperable wireless broadband network for public safety will ensure that first-responders have real-time access to critical information in an emergency.

Box 5.2. Broadband Technology Opportunities Program

The Broadband Technology Opportunities Program (BTOP) was created as part of the American Recovery and Reinvestment Act (ARRA) with the purpose of increasing broadband access and adoption in unserved and underserved areas; providing broadband training and support to schools, libraries, healthcare providers, and other organizations; improving broadband access to public safety agencies; and stimulating demand for broadband.

The Department of Commerce's National Telecommunications and Information Administration (NTIA) administers the BTOP program[1] and has invested $4 billion in 233 BTOP projects throughout all U.S. states, territories and the District of Columbia. Funds are awarded in three categories:

- Infrastructure. $3.5 billion to 123 projects for the construction of broadband networks
- Public Computer Centers. $201 million to 66 projects that will provide access to broadband, computer equipment and training to the public and vulnerable populations
- Sustainable Broadband Adoption. $251 million to 44 projects that promote broadband adoption, especially among vulnerable populations

Examples of BTOP projects include:[2]

The University Corporation for Advanced Internet Development (UCAID), also known as Internet^2, began upgrading its advanced middle-mile backbone network in March 2011.

This upgraded network will extend across 50 states and will enable high-speed broadband connectivity for up to 121,000 additional community anchor institutions such as schools and libraries. This large-scale, public-private partnership will interconnect more than 30 existing research and education networks, creating a dedicated fiber-optic backbone that will enable advanced broadband capabilities such as video multicasting, tele-medicine, distance learning, and other life-changing Internet-based applications. As of July 2011, more than 4,828 miles of its proposed 16,312 mile fiber network has been upgraded and activated with the entire network expected to be completed by early 2013.

The State Library of Louisiana's BTOP project, "Louisiana Libraries: Connecting People to Their Potential," has held more than 1,200 free digital literacy and software classes and supplied 640 laptop computers to libraries throughout the state by the end of 2011. The primary goal is to promote broadband use so that citizens may become comfortable and familiar with this technology. The State Library also seeks to improve the workforce skills of Louisiana citizens and provide a solid economic foundation for strengthening Louisiana communities.

[1] U.S. Department of Commerce, National Telecommunications and Information Administration 2010, *Expanding Broadband Access and Adoption in Communities Across AmericaOverview of Grant Awards*; www.ntia.doc.gov /files/ntia/publications/ntia_report_on_btop_12142010_0.pdf.

[2] Additional BTOP projects and grants awarded, by state, can be found at www2.ntia.doc.gov/.

The spectrum necessary for wireless communications, including broadband, has thus become an important resource to be integrated into the ICT infrastructure. Thus, a sensible policy for managing this spectrum is crucial if the United States is to improve its competitive position. The supply of spectrum is limited, however, and the rapid growth in demand driven by the high data consumption of smart phones and other mobile devices could result in a "spectrum crunch" in three to five years, severely inhibiting the development of next generation high-speed wireless technologies. Techniques such as improvements in spectral efficiency, increases in network density through cell site construction, and offloading traffic to wireline networks will not be sufficient to allow capacity to keep up with demand. In other words, wireless carriers will not be able to accommodate this surging demand without access to additional parts of the spectrum. It is vital that the government continue to address these spectrum challenges by reallocating spectrum from existing to more efficient uses.[18] One aspect of this reallocation is having Congress authorize the FCC to use auctions to reallocate spectrum from TV broadcasters to wireless broadband providers. This can also involve further research and development of technologies that enable more efficient use of spectrum.[19]

ICT also allows firms to collect huge amounts of data about their operations and use these data to improve productivity and increase innovation.[20] Big data—defined as "the large datasets generated from every customer interaction, every wired object, and every social network"[21]— as well as scientific and surveillance data, such as weather data from satellites, have the potential to provide information to improve greatly the efficacy of businesses, governments, health care and education.[22] The production of global data is expected to grow at an annual rate of 40 percent, providing a treasure trove to businesses and industry sectors that can effectively use these data to create new business models, new products and services, improve marketing strategies, and gain a competitive advantage.[23] Research has shown that firms that make effective use of these large datasets improve their productivity by 5 or 6 percent.[24]

However, to take advantage of these data capabilities, firms need access to broadband Internet networks with sufficient capacity to collect and distribute this information quickly and efficiently. They also need to have access to various computing resources such as servers, storage, applications, and services. These services increasingly are being made available through cloud computing, which provide companies with convenient, on-demand access to a shared pool of these necessary resources. With cloud computing, innovators and small businesses can offer new products and services to a global market with a very low investment of upfront resources, enabling them to compete with much larger providers. A final, closely related element of a sound ICT policy is an open Internet, one that protects consumers and innovators. Innovators need to be able to compete on their merits and not face anticompetitive barriers. Internet privacy is also crucial, and cybersecurity concerns need to be addressed.[25]

President Obama has pledged to preserve the free and open nature of the Internet to encourage innovation, protect consumer choice, and defend free speech. The Administration has created an Internet Policy Task Force to bring together industry, consumer groups, and policy experts to identify ways of ensuring that the Internet remains a reliable and trustworthy resource for consumers and businesses and has recently called for an Online Privacy Bill of

Rights. In July 2011, the Obama Administration joined with representatives from business, civil society, and Internet technical communities from 34 countries to reaffirm the importance of Internet policy principles that have enabled the open Internet to flourish with innovation and human connections beyond our wildest expectations.

Americans deserve an Internet that is safe and secure, so they can shop, bank, communicate, and learn online without fear their accounts will be hacked or their identities stolen. President Obama has declared that the "cyber threat is one of the most serious economic and national security challenges we face as a nation" and that "America's economic prosperity in the 21st century will depend on cybersecurity." To help the country meet this challenge and to ensure the Internet can continue as an engine of growth and prosperity, the Administration is implementing the National Strategy for Trusted Identities in Cyberspace. The Administration also released the International Strategy for Cyberspace to promote the free flow of information, the security and privacy of data, and the integrity of the interconnected networks, which are all essential to American and global economic prosperity and security.

The Obama Administration has made cybersecurity at Federal departments and agencies a priority and it is moving forward on the government's implementation of the Federal Information Security Management Act (FISMA). In addition, the Obama Administration is working on the development of the National Initiative for Cybersecurity Education, which will establish an operational, sustainable, and continually improving public cybersecurity education program to promote sound cyber practices within the United States.

Satellites

An increasingly important part of the infrastructure of the United States can be found orbiting the earth in the form of our satellite system. Satellites are used for many crucial tasks, such as improving weather forecasting so that businesses have the most reliable and up-to-date information for planning purposes. To meet these goals, the Administration is acquiring and operating the satellites needed to support weather forecasting, climate monitoring, and ocean and coastal observations. The Administration also plans to launch a new generation of global positioning satellites and services in order to implement advanced navigation and timing applications that can support innovation in many sectors, including agriculture, communications, air travel, and highway safety.[26] For example, as discussed below, global positioning satellites will be a key part of a new, Next Generation (NextGen) air traffic control system.

Finally, the Administration is committed to maintaining international partnerships to further improve on our capabilities in these areas.[27]

Ensuring the United States' 21st Century Infrastructure is Sound

Our nation faces significant challenges in rebuilding its infrastructure and wise investments in these areas have to be made. Below are several specific policy proposals that will address the problems discussed above.

NextGen Air Traffic Control System

Since there are often constraints on increasing the number of airports and runways, another way to increase the capacity of the existing physical air traffic infrastructure is by making it more efficient; that is, by increasing the number of flights that can be handled on existing runways. This can be achieved through the adoption of the NextGen air traffic control system, a comprehensive overhaul of the existing system.[28] The current system is based on radar, whereas NextGen will employ global positioning systems and new ground-based and airborne technologies to deliver new communications, navigation, surveillance, and information management capabilities. As a result of this more accurate information, NextGen will allow more aircrafts to fly safely closer together on more direct routes, reducing fuel usage, noise, and flight and ground delays by 35 percent. This reduction in delays will translate to $23 billion in cumulative benefits to the Federal Aviation Administration, airlines, and travelers. In addition, aviation fuel usage would decline by 1.4 billion gallons and carbon dioxide emissions would fall by 14 million tons.[29]

While NextGen's overall benefits are compelling, adoption of NextGen faces significant hurdles, such as the merging of varying technologies and interfaces, maintaining and retaining a skilled FAA workforce, and keeping to the current rollout schedule given the current budget climate. FAA authorization has been temporarily extended 22 times since the previous long-term FAA authorization expired in 2007. The current extension is set to expire on January 31, 2012.[30]

Wireless Communications

The Obama Administration has made it a priority to improve the wireless broadband infrastructure in the United States. A "National Wireless Initiative" was announced in February 2011 with the stated goals of doubling the amount of spectrum available for wireless broadband services and helping rural areas gain access to wireless broadband services through reform of the FCC Universal Service Fund so that it focuses more on wireless service rather than traditional phone service.

Cloud Computing

Various initiatives are underway to help companies process large amounts of data through cloud computing. For example, NSF has been working with Microsoft to provide free access to Microsoft's cloud products to researchers who either want to investigate further the possible ways in which cloud computing can be used or have projects related to biotechnology or other industries that require large amounts of data storage. NASA has created a cloud computing platform called "Nebula" that allows NASA scientists to share large, complex data sets with research partners and the general public.

Open Innovation Strategy

Data is a critical component in information technology and the Administration's open innovation strategy is meant to increase access to data to help spur innovation. This strategy incorporates four major components:

- **Improve Access to Government Data.** On his first full day in office, President Obama signed a memorandum on transparency and open government. Part of this initiative was the launch of data.gov, a platform that provides public access to high value, machine readable datasets, now numbering in excess of 390,000.
- **Encourage Market Transparency.** A transparent "marketplace" will lower barriers to entry and unleash the creativity of entrepreneurs to compete in the development of new consumer-oriented products and services. The Obama Administration is working with the health, energy, and education sectors (among others) to simplify access to high value data by, for example, encouraging the creation of standards.
- **Cultivate Innovation Communities.** Bringing together communities of innovators will help spur innovation. To support these communities, the Administration has partnered with organizations to inspire participation in innovative activities through the use of challenges and prizes.
- **Create Capacity for Innovation.** To manage these policy tools, the Administration has actively recruited a group of technology and innovation leaders with direct reporting relationships to the Cabinet

Secretaries. In turn, these leaders are recruiting three to five person "innovation teams" to tackle an identified problem with rapid results.

One example of the implementation of this strategy is HealthCare.gov, which launched on July 1, 2010. Its marquee attraction, the Insurance Finder, asks the user a few simple questions and then produces a customized menu of insurance choices that draws from a huge inventory of offerings from over 1,000 insurance companies and every major public health coverage program in the country.

Smart Grids

As mentioned above, the Federal government recently released a policy framework to help promote investment, innovation, and job growth in the area of Smart Grids. The main elements of this framework include enabling cost-effective Smart Grid Investments; ensuring that there are appropriate standards in place so that grids can interconnect with each other; giving consumers the information needed to save energy; and ensuring that Smart Grids are protected from cyber attacks and, in the event of such an attack, the systems can quickly recover.

References

Brown, Brad, Michael Chui, and James Manyika. 2011 "Are You Ready for the Era of Big Data." *McKinsey Quar terly* (October): 5–13.

Brynjolfsson, Erik, Lorin M. Hitt, and Heekyung Hellen Kim. 2011. "Strength in Numbers: How Does Data-Driven Decisionmaking Affect Firm Performance?" Working Paper Series. (April):1–28; papers.ssrn.com/sol3 /papers.cfm?abstract_id=1819486.

Council of Economic Advisers. 2011. *The Economic Report of the President.* Washington, DC: U.S. Government Printing Office, February.

Council on Environmental Quality. 2011. "Administration Announces Grid Modernization Initiatives to Foster a Clean Energy Economy and Spur Innovation." Press Release; www.whitehouse.gov/administration/eop/ceq/ Press_Releases/June_013_2011.

Economics and Statistics Administration and National Telecommunications and Information Administration. 2011. *Exploring the Digital Nation: Computer and Internet Use at Home.* Washington, DC: U.S. Department of Commerce, November; www.esa.doc.gov/sites/default/files /reports /documents/exploringthedigitalnationcomputerandinternetuseathome.pdf.

Edison Electric Institute. 2011. "Electric Utilities: The Power of Opportunities." *2011 Wall Street Briefing* (February 9): 1–9; www.eei.org /ourissues/finance/Documents/Wall_Street_Briefing_2011.pdf.

Federal Aviation Administration. 2011. *NextGen Implementation Plan.* Washington, DC: NextGen Integration and Implementation Office, March: 5; www.faa.gov/nextgen/media/ng2011_implementation_plan.pdf.

Federal Communications Commission. 2011a. *Connecting America: The National Broadband Plan.* Washington, DC: March; download. broadband.gov/plan/national-broadband-plan-chapter-1-introduction.pdf.

Federal Communications Commission. 2011b. *Connect America Fund & Intercarrier Compensation Reform Order and FNPRM.* Washington, DC: October; e-ratecentral.com/archive/Bulletins/2011_10_27.pdf.

Katz, Bruce, and Robert Puentes. 2008. "America's Infrastructure: Ramping Up or Crashing Down." Washington, DC: The Brookings Institution, January; www.brookings.edu/papers/2008/01_infrastructure_katz_puentes.aspx.

Manyika, James and Charles Roxburgh. 2011. The Great Transformer: *The Impact of the Internet on Economic Growth and Prosperity.* McKinsey Global Institute. October; www.mckinsey.com/Insights/MGI/Research/ Technology_and_Innovation/The_great_transformer.

Manyika, James, Michael Chui, Jacques Bughin, Brad Brown, Richard Dobbs, Charles Roxburgh, and Angela Hung Byers. 2011. *Big Data: The Next Frontier for Innovation, Competition, and Productivity.* McKinsey Global Institute. May; www.mckinsey.com/Insights/MGI/Research/Technology _and_Innovation/Big_data_The_next_ frontier_for_innovation.

National Economic Council, Council of Economic Advisers, and Office of Science and Technology Policy. 2011. *A Strategy for American Innovation: Securing Our Economic Growth and Prosperity.* Washington, DC, February; www.whitehouse.gov/sites/default/files/uploads/Innovation Strategy.pdf.

National Science and Technology Council. 2011. *A Policy Framework for the 21st Century Grid: Enabling Our Secure Energy Future.* Washington, DC: Executive Office of the President, June; www.whitehouse.gov/sites/default /files/microsites/ostp/nstc-smart-grid-june2011.pdf.

National Telecommunications and Information Administration. 2011. *Second Interim Progress Report on the Ten-Year Plan and Timetable.*

Washington, DC: October; www.ntia.doc.gov/files/ntia/publications /second_interim_ progress_report on the ten year_plan_and_timetable.pdf.

Pélissié du Rausas, Matthieu, James Manyika, Eric Hazan, Jacques Bughin, Michael Chui, and Rémi Said. 2011. *Internet Matters: The Net's Sweeping Impact on Growth, Jobs and Prosperity.* McKinsey Global Institute. May; www.mckinsey.com/Insights/MGI/Research/Technology_and_Innovation/ Internet_matters.

Reilly, Sean. 2011. "Congress Extends Funding for FAA through January, Prevents Furloughs," *Federal Times,* updated September 16. Accessed December 4, 2011; www.federaltimes.com/article/20110916/AGENCY 02/ 109160302/.

U.S. Department of Energy. 2010. "Statement of Secretary Steven Chu Before the Senate Committee on Energy and Natural Resources FY 2011 Budget Hearing." February 4; energy.gov/sites/prod/files/ciprod/documents/2-4-10_Final_Testimony_%28Chu%29.pdf.

Waverman, Leonard, Kalyan Dasgupts, and Janne Rajala. 2011. *Connectivity Scorecard 2011.* Report. Emeryville, CA: Berkeley Research Group. May; www.connectivityscorecard.org/images/uploads/media TheConnectivity Report2011.pdf.

White House. 2010. "National Space Policy of the United States of America." June 28; www.whitehouse.gov/ sites/default/files/national_space_policy_ 6–28-10.pdf.

Revitalizing Manufacturing

"Build it here. Sell it everywhere."
— *John E. Bryson, Secretary of Commerce, December 15, 2011*

A Strong Manufacturing Sector is Uniquely Important to the U.S. Economy

A flourishing manufacturing sector in the United States is crucial to its future competitive strength. Throughout its history, manufacturing has been a source of prosperity, innovation, and pride for the United States. Manufacturing pays higher than average wages, provides the bulk of U.S. exports, contributes substantially to U.S. R&D, and protects national security.

Manufacturing remains a vital part of the U.S. economy. In 2009, manufacturing made up 11.2 percent of gross domestic product (GDP)[1] and 9.1 percent of total U.S. employment,[2] directly employing almost 12 million workers. This sector also has indirect employment effects on other sectors of the U.S. economy when it purchases inputs for production such as raw materials (such as from the agricultural and mining sectors), buildings (from the construction and real estate sectors), and services (including warehousing and transportation; professional, scientific, and technical services; and financial services). In these ways, manufacturing supports millions of additional supply chain jobs across the economy. In addition, many of the jobs provided by this sector are high quality. Total hourly compensation in the manufacturing sector is, on average, 22 percent higher than that in the services sector and about 91 percent of factory workers have employer-provided benefits compared to about 71 percent of workers across all private sector firms.[3] Manufacturing is also the largest contributor to U.S. exports. In 2010, the United States exported over $1.1 trillion of manufactured goods, which accounted for 86 percent of all U.S. goods exports and 60 percent of U.S. total exports (see figure 6.1). In order to support millions more jobs, President Obama's National Export Initiative set the ambitious goal of doubling U.S. exports by the end of 2014. Moreover, the United States runs a trade surplus in the services sector, a surplus that has tripled since 2003[4]; however, though the services sector will continue to be important, increases in services alone will not likely double U.S. exports by 2014. Indeed, without a strong manufacturing sector, the U.S. trade surplus in services may erode (see box 6.1). A strong manufacturing sector is also crucial because successful innovation in many sectors is closely linked to the ability to manufacture products as innovative methods and ideas are generated and perfected through the process of making things. In the recent *Report to the President on Ensuring American Leadership in Advanced Manufacturing,*[5] the President's Council of Advisors on Science and Technology (PCAST) and the President's Innovation and Technology Advisory Committee (PITAC) emphasize the critical importance of advanced manufacturing in driving knowledge production and innovation in the United States. The PCAST researched the current state of manufacturing and concluded that U.S. leadership in manufacturing is declining and that this is detrimental to the well-being of the nation overall. Manufacturing companies in the United States are responsible for over two-thirds of the industrial R&D[6] and employ the majority of domestic scientists and engineers.[7] Furthermore, manufacturing R&D is the dominant source of innovative new service-sector technologies,8 hence its benefits reach beyond the manufacturing arena.

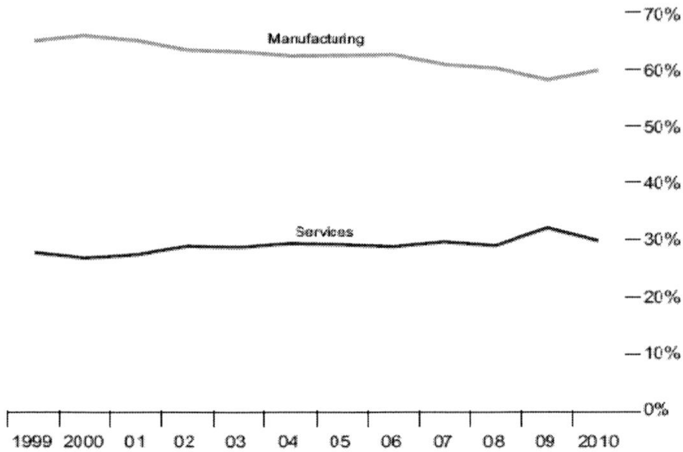

Source: Bureau of Economic Analysis and Census, U.S. International Trade in Goods and Services; excluding Agriculture and Non-agriculture/Non-Manufacturing goods.

Figure 6.1. U.S. Exports by Sector, Share of Total, 1991–2010.

Box 6.1. Tradable Sectors: A Source of Good Jobs

Manufacturing is generally a "tradable" sector; that is, an activity that can be "transacted across distances" thus making it vulnerable to import competition. Since the jobs in manufacturing generally pay well, the loss of these jobs due to import competition can have severe negative effects on the well-being of the U.S. workforce. Service activities at one time were not considered tradable, but some service industries have become an important and expanding component of U.S. trade. Tradable service jobs are also high-quality, with higher education and wage levels than jobs in non-tradable services.[1] Given the recent decline in U.S. manufacturing, in part due to off-shoring, a concern is whether this will happen to the service sector. In fact, it could be argued that many of the current tradable services exist because various firms had a strong manufacturing capability that also provided a source of highly-trained engineers and technical staff that could export these services.

Without a core manufacturing capability feeding that engineering base it could be argued that long-term growth in tradable services is not sustainable. On the other hand, rather than lose jobs, the comparative advantage of the United States in high-skill, high-wage service jobs such as engineering and business services points to potential opportunities to expand services exports and increase jobs in these areas.

[1] Jensen, J. Bradford. August 2011. "Global Trade in Services: Fear, Facts, and Offshoring." Peterson Institute for International Economics, Washington, DC. bookstore.piie.com/book-store//6017.html.

The colocation of manufacturing, research, and other sectors can also be important. In its recent report the PCAST states: "Proximity is important in fostering innovation. When different aspects of manufacturing—from R&D to production to customer delivery—are located in the same region, they breed efficiencies in knowledge transfer that allow new technologies to develop and businesses to innovate."[9] Thus, even if R&D facilities are kept in the United States, the relocation of manufacturing facilities overseas may limit the United States' ability to innovate. Finally, an innovative and secure domestic manufacturing base is critical to national security. An inability to produce domestically the advanced defense systems of the modern military would put the national security of the United States at risk. As its military comes to rely more heavily on complex and advanced technology systems, it is important that the United States retain the manufacturing capacity and knowledge necessary to produce these goods. Our continued security not only rests on the ability to produce military products, but we must also consider how the sourcing of all critical infrastructure components, from communications equipment to power generation, affects our ability to protect against potentially catastrophic supply chain disruptions.

The Current State of U.S. Manufacturing: A Crossroads for American Competitiveness

While manufacturing continues to play a vital role in the U.S. economy and provides millions of American jobs, the U.S. manufacturing sector has faced significant challenges in recent decades. As a fraction of U.S. GDP, manufacturing declined from 27 percent in 1957 to about 11 percent by 2009[10] (see figure 6.2).

Since 2000, increases in the manufacturing of high tech equipment (semiconductors and related components, computers, and communications equipment) have hidden a slight decline in output of all other manufacturing sectors (see figure 6.3).[11] Manufacturing employment has seen dramatic declines; in the last decade alone, employment levels in manufacturing have declined by about a third (see figure 6.4), and the impact of this decline in manufacturing employment has been felt in many states across the country, with several states experiencing near collapse of their manufacturing sectors (see figure 6.5).

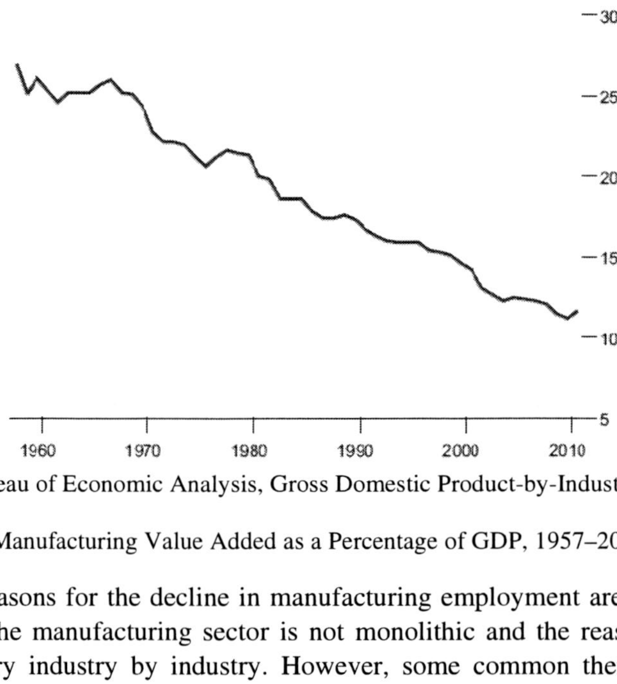

Source: Bureau of Economic Analysis, Gross Domestic Product-by-Industry Data.

Figure 6.2. Manufacturing Value Added as a Percentage of GDP, 1957–2010.

The reasons for the decline in manufacturing employment are varied and complex; the manufacturing sector is not monolithic and the reasons for the decline vary industry by industry. However, some common themes can be discerned. One likely factor is the large improvement in productivity in manufacturing. Between 1987 and 2010, labor productivity in manufacturing rose at a 3.4 percent annual rate, almost 50 percent higher than the 2.3 percent annual rate in the entire nonfarm business sector.[12] Though this increased productivity is critical in terms of maintaining the competitiveness of the manufacturing sector, it also has meant that companies can now do the same work with fewer workers and partially explains the drop in employment over the past two decades. However, given that employment in manufacturing held relatively steady during the 1990s while productivity was still high, other factors must also play a role in the decline of manufacturing since 2000. Further, a large portion of the overall gains in productivity are attributable to the production of computer and electronics products, so productivity gains are less likely to explain employment declines in other industries. In addition, this overall improvement in productivity may be slightly overstated due to the fact that low-cost foreign inputs are not adequately captured in existing price indices.[13]

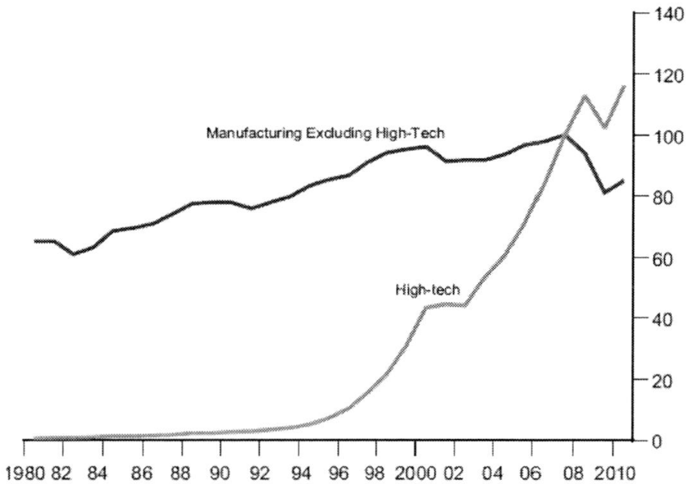

Source: Federal Reserve and Haver Analytics.
Note: 2007=100, Indexed.

Figure 6.3. High-Tech Manufacturing Production Versus Manufacturing Production Excluding High-Tech, 1980–2010.

Another factor in the employment losses, particularly in some less efficient industries, is greater competition from low-wage countries, leading to the off-shoring of low-skilled jobs to lower cost locations. For example, one study has shown that between one-quarter to more than one-half of the lost manufacturing jobs in the 2000s was the result of import competition from China.[14] While there has been an overall decline in manufacturing employment, as stated above, there is evidence that the extent to which employment has fallen varies according to the amount of competition an industry faces from imports from low-wage countries. In fact, between 1972 and 2001, industries that faced the most import competition from low-wage countries saw an average decade-long decline in employment of 12.8 percent, while industries that faced little low-wage import competition saw an increase in average decade-long employment of 2.3 percent.[15]

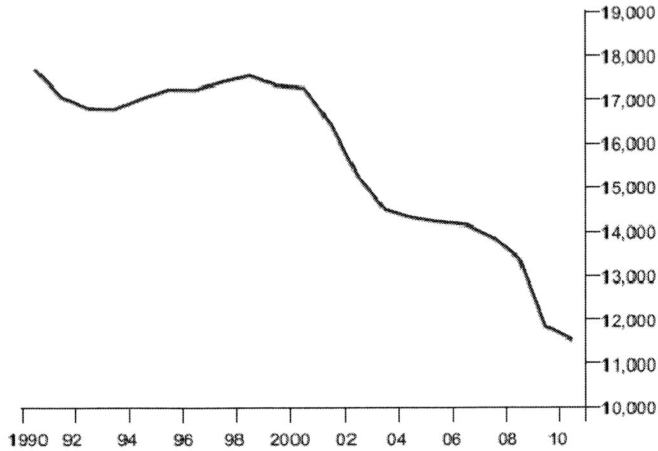

Source: Bureau of Labor Statistics, Employment Nonfarm payrolls.

Figure 6.4. Manufacturing Employment, (Thousands) 1990–2010.

Note: Percent Change in Manufacturing Employment by state, 1990 (first year of data) to 2007.

Source: Bureau of Labor Statistics, Economics and Statistics Administration calculations.

Figure 6.5. Percent Change in Manufacturing Employment, 1990–2007.

While much has been written about the decline in jobs for unskilled labor within traditional manufacturing, this is only part of the story. The United States is also losing ground in the manufacture of high-tech goods that require skilled labor (see figure 6.6). PCAST notes, the United States has "not simply lost low-value jobs, such as assembly, in the high-tech sector, but also sophisticated engineering and advanced manufacturing activities."[16] This could be due to various factors; other countries may have relatively more skilled labor, may produce higher quality products, or have better customer service. The relative strength of the U.S. dollar can also play a role.

While some might suggest that an advanced country, such as the United States, will inevitably lose manufacturing share as the country shifts towards a more services-oriented economy, this is not a foregone conclusion. As the *Economic Report of the President* notes, "experience from other high-income countries shows that a shift in the world share of exported goods does not mean a shift entirely out of manufacturing and into a service-only economy. Germany, the second place goods exporter, maintains a substantial share of manufacturing in its economy and exports many of these products...manufacturing, especially of complex products, continues to play a substantial role in advanced economies."[17]

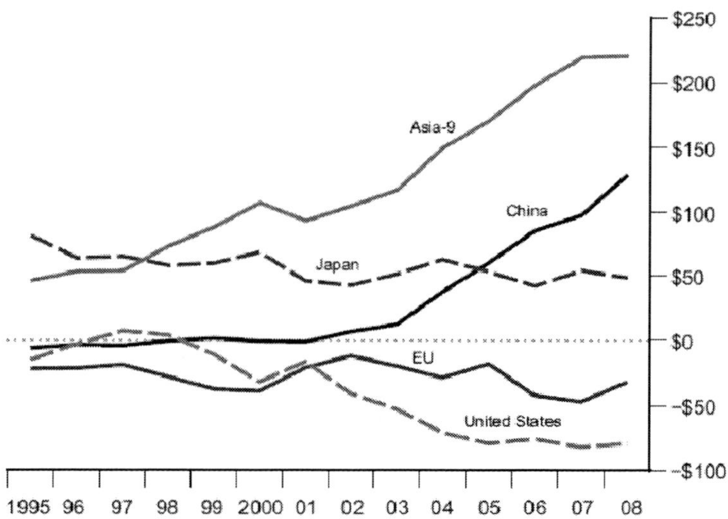

Source: IHS Global Insight, World Trade Service database (15 July 2009)*
Note: Billions of dollars.

Figure 6.6. Trade Balance of High-Tech Goods, 1995–2008.

Economic Rationales for Federal Government Support for U.S. Manufacturing

An overarching U. S. manufacturing innovation policy should invest to overcome market failures and to ensure technology-based enterprises have the infrastructure needed to be successful. The Federal government can help facilitate this by supporting research programs in new technologies; supporting the creation and dissemination of powerful design methodologies that dramatically expand the ability of entrepreneurs to design products and processes which any given entrepreneur may not have the incentive to invest in on its own; and investing in shared technology infrastructure that would help U.S. companies improve their manufacturing.

The manufactured goods market is global and companies from more and more countries are participating. It is not an option or desired outcome for the United States to close its borders to goods produced abroad; however, we must be conscious of the impacts that our government's actions and those of our trading partners have on the competitiveness of this sector. (The ways in which foreign governments support industry are discussed more in Chapter 7).

The Federal government has historically played an important supportive role in the manufacturing sector. As *A Framework for Revitalizing American Manufacturing* noted, "the key to success (in manufacturing) lies in American workers, businesses, and entrepreneurs—but the federal government can play a supportive role in providing a new foundation for American manufacturing."[18] Just as there is no single explanation for why manufacturing has declined in the United States, no one policy prescription will reverse the decline.

Successful manufacturing policy actions must reflect the diversity of the manufacturing sector while not creating an industrial policy that inefficiently seeks to pick winners and losers.

Longstanding Federal Government Support for U.S. Manufacturing

Trade Policies

The United States works to open markets for U.S. goods and services through free trade agreements and other activities. The Federal government also takes steps to enforce existing trade rules within the World Trade Organization framework. Unfair foreign pricing and government subsidies distort the free flow of goods and adversely affect some American businesses in the global marketplace. Free trade must be premised on fair trade.

Investments in Research and Development Infrastructure

As noted in Chapter 3, Federal support for R&D provides a vital and necessary public good that individual private companies may be unwilling or unable to undertake. Federal support for R&D, particularly support for long-term basic research, has helped the advancement of important innovative technologies that have then spawned many successful companies and even entire industries (for an example, see box 6.2).

In their *Report to the President,* PCAST notes, "The Federal Government has historically made visionary investments that have facilitated the birth of new technology-based industries and strengthened the development of existing industries. These investments have paid enormous financial and social returns to the Nation."[19]

The Federal government supports R&D through agencies such as NSF, DARPA, NIST, and the DOE's Office of Energy Efficiency and Renewable Energy (see box 6.3 for a detailed description of NIST's manufacturing-related activities and see box 6.4 for an example of a company that has benefitted from multiple Federal programs).

The Federal government also has played a role by helping to fund large-scale research labs as part of public-private partnerships. As noted by a recent PCAST report, in the past the Federal government "funded in part the major corporate laboratories that laid the foundations for the U.S. economic leadership and inno vation in the 20th century, including Xerox PARC, RCA David Sarnoff Research Center, and AT&T Bell Labs."[21]

Box 6.2. A123 Systems: Supporting the Future of the Auto Industry

In 2001 Professor Yet-Ming Chiang of the Massachusetts Institute of Technology co-founded A123 Systems, a producer of rechargeable lithium-ion batteries that power hybrid and electric vehicles and other technologies. The firm's foundation was enabled by a Small Business Innovation Research grant from the Department of Energy. The firm subsequently raised more than $300 million in capital from investors like Sequoia Capital and corporations like GE and Motorola. It also received a $5 million loan from the Massachusetts Clean Energy Center.[1] It went public in September 2009.

In February 2010, it embarked on a $6.0 million research program, with funding from the NIST Technology Innovation Program (TIP), to develop a new nanocomposite material for lithium ion battery electrodes together with improved manufacturing process technologies to enable both significantly improved battery performance and lower manufacturing costs.

With help from the Department of Energy it opened a manufacturing plant in Michigan in September 2010.2 Today A123 Systems employs approximately 1,700 people.

[1] National Economic Council and the Office of Science and Technology Policy. *A Strategy for American Innovation: Securing Our Economic Growth and Prosperity.* February 2011, 58. www.whitehouse.gov/sites/default/files /uploads/InnovationStrategy.pdf.

[2] The Science Coalition, *Sparking Economic Growth: How federally funded university research creates innovation, new companies and jobs.* April 2010, 35. www.sciencecoalition.org/successstories/resources/pdf/Sparking %20 Economic%20Growth%20Full%20Report%20FINAL%204-5-10.pdf.

Box 6.3. The National Institute of Standards and Technology (NIST)

The National Institute of Standards and Technology (NIST) helps manufacturers of all kinds—from shipbuilders to semiconductor makers—streamline their operations, improve quality, reduce environmental impacts, and develop innovative products and processes.

The NIST Laboratory programs also provide critical support for manufacturers through research, standards activities, and the development and delivery of measurement services.

Efforts are underway in partnership with industry and academia, to produce measurement technologies, standards, and services in areas including nano- and biomanufacturing, advanced robotics, additive manufacturing, cyberphysical systems, advanced materials development, and a number of other areas that will broadly impact technologies that are critical to advanced manufacturing across industry sectors.

NIST is committed to advancing the Administration's vision for advanced manufacturing and will continue to provide:

- Unique and enabling measurements to industry, particularly in support of emerging technologies. In the area of advanced materials NIST is working to develop modeling and characterization tools that will help reduce materials design time from the current 10 year timeframe to a timescale more compatible with the average 18 month product development cycle.
- Support to strategic standards development and adoption. In the area of robotics NIST is working to provide the measurement framework that will support the adoption of standards to enable safer, closer proximity humanrobot interactions on the factory floor.
- Support to technology transfer and commercialization of technology. In the area of nanomanufacturing the NIST nano-fabrication facility provides a key facility for users to test new manufacturing methods and techniques that can help speed the introduction of new nanomaterials into new products.

In addition, NIST's Hollings Manufacturing Extension Partnership (MEP) and its nationwide network of centers have a proven track record of helping manufacturers. According to NIST research "every $1 of Federal investment in MEP generates $32 of return in sales growth, a total of $3.6 billion in new sales nationally."20 MEP centers offer access to market intelligence, trends, and data about manufacturing; outreach assistance to existing manufacturing firms in the region to get them involved in cluster initiatives (particularly small and medium sized manufacturers); technical assistance to companies in targeted clusters to enhance their competitiveness and accelerate growth opportunities (technology development, sustainability, etc.) leading to job creation; and tracking of performance measures (e.g., jobs created/retained, cost savings, new sales, new investments).

Box 6.4. West Paw Designs: Sustainable Manufacturing in Montana

West Paw Design is a small manufacturer based in Bozeman, MT that makes pet toys and beds.[1] West Paw uses IntelliLoft—a fiber created from 100 percent post-consumer recycled plastic soda bottles—to fill their stuffed pet beds and toys.

Since 2006, the company has helped divert more than 5 million plastic bottles from landfills through this practice.

West Paw has taken advantage of Federal programs and services for small businesses, for sustainable manufacturing, and for exporting. They've been able to utilize the variety of services available to them, including SBA loans through the Recovery act, the Hollings Manufacturing Extension Partnership program at NIST, and some export assistance from the Department of Commerce's U.S. Commercial Service.[2] In 2010 West Paw doubled the size of its manufacturing facility by focusing on a green line of products and by looking to the international marketplace, with the help of various Federal government programs and services.

[1] West Paw Design, "The West Paw Design Story." Accessed November 15, 2011 www.westpawdesign.com/articles/-west-paw-story/west-paw-design-story.

[2] Williams, Spencer. Invited Testimony before the United States Senate Committee on Finance. February 23, 2010 finance.senate.gov/imo /media/doc/022310swtest.pdf.

Investments in Education

As outlined in Chapter 4, investments in education, particularly STEM education, are critical to the future competitiveness of the United States. This is especially true for modern manufacturing, which requires an increasingly skilled workforce. Just as the manufacturing sector today is diverse and not a monolithic set of fac tories banging out widgets, today's manufacturing workforce is diverse, with a wide range of skills. The share of manufacturing employment accounted for by those with at least some college education has been increasing over time and ex ceeded half of the overall manufacturing labor force during the last few years (see figure 6.7).

Community colleges are educating many of these higher skilled manufacturing workers either as a continuation of their formal K–12 education or as part of the workforce development system. The United States' public, 2year college system has more than 7.1 million students enrolled and awards 790,000 associate degrees annually.[22] Moreover, community colleges award huge numbers of non-degree certificates in specific scientific, technical, and computing skills. In addition, many of these institutions offer contract training for the public sector and employers, providing multiple opportunities for students and workers to gain skills that can facilitate their job search or allow them to become more productive in their current jobs. Students have been flocking to public 2-year colleges, with enrollment up by 75 percent between 1979 and 2009, and by 12 percent between December 2007 and June 2009.[23]

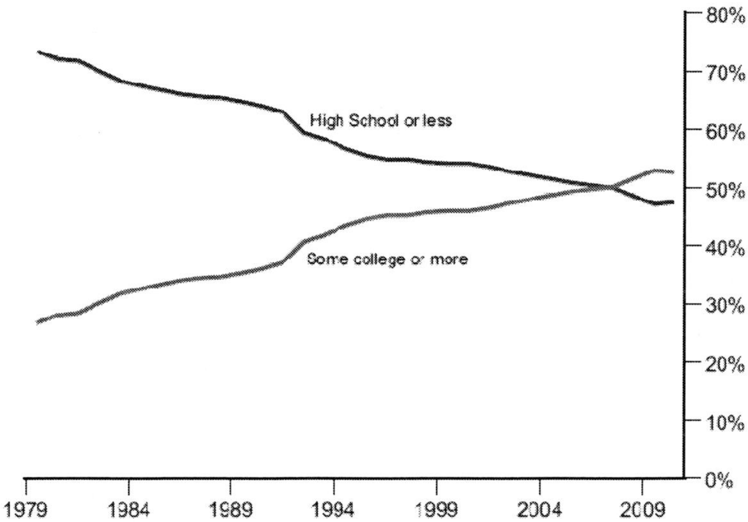

Source: CPS public use Merged Outgoing Rotation Group files from NBER.

Figure 6.7. Manufacturing Employment by Education Level, 1979–2010.

Employment projections through 2018 show that jobs that require at least some postsecondary education will be growing faster than those that require workers with just a high school diploma or less; however, the fastest growth will be in jobs for which an associate degree is the best pathway of entry.[24] Community colleges are also a needed nexus between industry and higher education, providing education in academic fields, including STEM, combined with vocational studies (see box 6.5 for an example of private-public partnership at community colleges).

Community colleges depend much more than 4-year public universities on state and local government appropriations. In the 2008–2009 school year, 47 percent of total revenues of public 2-year schools came from these appropriations, compared with 24 percent for public 4-year schools.[25] Given their dependence on state and local budgets, community colleges are especially vulnerable to government cutbacks. The Obama Administration recognized early on the essential role played by community colleges, and the $2 billion Health Care Reform Act investment in community colleges is one essential and timely investment that will help strengthen not just the colleges themselves, but also their ties to local industries. While community colleges by definition operate at a local level, these needed Federal government investments support workers, their communities, and the nation's industrial base.

Box 6.5. 10,000 Small Businesses

This Goldman Sachs initiative is a $500 million, five-year program that aims to unlock the growth and job-creation potential of 10,000 small businesses across the United States. It provides access to business education, mentors, networks and financial capital. The program is anchored at local community colleges. At year end 2011, the program was operating in New York, Los Angeles, Chicago, New Orleans, Houston, and Long Beach, CA, and focused on historically underserved communities.

Economic development experts believe that a combination of education, capital and support services best address the barriers to growth for small businesses. The current environment of fiscal austerity is notably impeding the budgets of many public post-secondary school programs, including community colleges that often provide support to new business owners and vocational training to others.

The Goldman Sachs program has thus far targeted disadvantaged urban areas. The board of 10,000 Small Businesses, which includes Warren Buffet and Professor Michael Porter, has laid out the mandate for the initiative which is to meet the vital need for training, tools and relationships to help local entrepreneurs create a self-reinforcing cycle of economic opportunity.

Investments in Transportation, Energy, and Communications Infrastructure

Finally, the Federal government can support American manufacturers by investing in a 21st century infrastructure, as outlined in Chapter 5. This is because the "cost to move goods from one factory to another and to their final destination, the cost to move energy from where it is created to where it is used, the cost of moving people and the cost to transport information are all significant factors in the manufacturing process" notes the Administration's *A Framework for Revitalizing American Manufacturing*.[26] Also, PCAST notes that small-and medium-sized firms "would benefit from readily accessible shared infrastructure, providing both equipment and expertise. Infrastructure currently provided at Federal laboratories, for example, for the fabrication of micro-electromechanical systems, has allowed for new products to be developed."[27]

In addition to programs that are strictly Federal, partnerships and coordination with governments at the state and local level have also proved effective. For example, Commerce's NIST MEP, along with the Economic Development Administration (EDA), recently partnered with the National Governors Association (NGA) to launch a Policy Academy that will encourage the growth of advanced manufacturing industries (see box 6.6).

These examples clearly illustrate the important role of the Federal government in supporting U.S. manufacturing. This support has been important in the past and will likely be even more important in the increasingly competitive marketplace of the future.

Box 6.6. EDA, NIST, NGA Collaborate to Form a Policy Academy

The U.S. Commerce Department's NIST Manufacturing Extension Partnership (MEP), in collaboration with EDA, have partnered with the National Governors Association (NGA) to launch a Policy Academy to encourage coordination amongst stakeholders in both Federal and state government along with leaders in industry and academia, to spur the growth of advanced manufacturing industries and support American jobs.

The states will receive guidance and technical assistance from NGA staff, experts from MEP, EDA and the State Science and Technology Institute, as well as consultants from the private sector, research organizations and academia. Colorado, Connecticut, Illinois, Kansas, Massachusetts, New York and Pennsylvania have been selected to participate, and the strategies and policies that are developed at the Policy Academy are intended to benefit all states.

For more information, visit www.nga.org/cms/center/ehsw.

Federal Initiatives to Revive Manufacturing

Many initiatives are underway to revitalize the U.S. manufacturing sector. They include:

- *The White House Office of Manufacturing Policy.* To improve the coordination of manufacturing policy across the Federal government, President Obama announced on December 12, 2011 that Commerce Secretary John Bryson and National Economic Council Director Gene Sperling will be co-chairs of the White House Office of Manufacturing Policy. The office will convene cabinet-level meetings to implement and coordinate priority manufacturing initiatives.
- *The Advanced Manufacturing Partnership (AMP).* Launched in June 2011, AMP identifies opportunities for industry, academia, and government to collaborate in order to accelerate the development and deployment of emerging technologies with the potential to transform and reinvigorate advanced manufacturing in the United States.
    - The AMP Steering Committee (AMP-SC) is co-chaired by Susan Hockfield of the Massachusetts Institute of Technology and Andrew Liveris of Dow Chemical and includes leading experts from industry and academia, including CEOs of major manufacturing firms and presidents of leading universities. The AMP-SC conducted four regional meetings from October to December of 2011, and will be issuing a final report in the spring of 2012.
    - In addition, to support the rapidly advancing work of the AMP, the Administration is establishing a National Program Office (NPO) that will reside at Commerce's NIST and will be staffed by a broad representation from several key Federal agencies involved in U.S. manufacturing in order to provide a coordinated "whole-of-government" response. The AMP NPO will support the ongoing work of the AMP partners, support interagency coordination of advanced manufacturing programs, and provide a link to the growing private sector partnerships between manufacturers, universities, state and local governments, and other manufacturing-related organizations.
- *The Materials Genome Initiative.* This program modeled on the Human Genome project that deciphered the building blocks of human genetics, will speed understanding of fundamental issues related to materials science by investing in research, training and infrastructure to enable U.S. companies to discover, develop, manufacture, and deploy advanced materials. For example, the initiative will fund various computational tools and software to help understand the

properties of these materials and open standards and databases to help facilitate the sharing of knowledge.

- *SelectUSA* was established by Executive Order on June 15, 2011. It is the first Federal effort designed with executive authority to support foreign and domestic business investment in the United States. It showcases the United States as the world's premier business location, complementing the activities of states and regions—the primary drivers of economic development—to spur economic growth and job creation. SelectUSA coordinates existing resources and functions across all Federal agencies that have operations relevant to business investment decisions.

SelectUSA encourages business investment by conducting four critical, inherently governmental functions:

- **Outreach and engagement.** Leading and coordinating outreach and engagement by the Federal government to promote the United States as the best market for business operations in the world;
- **Ombudsman.** Serving as ombudsman to facilitate the resolution of specific issues involving Federal programs or activities related to pending investments and addressing the Federal regulatory climate through an interagency investment facilitation task force;
- **Information clearing house.** Providing information to firms regarding items such as Federal programs and incentives available to investors and state and local economic development points of contact; and,
- **Policy advisement and engagement.** Advising the White House, Federal agencies, and the U.S. economic development community on business investment policy issues based on feedback, solicited and unsolicited, that is received from investors and stakeholders.

- *New Federal support for R&D.* Initiatives the Obama Administration is championing include funding for DOE to support R&D in areas such as flexible electronics for components like batteries and solar cells and ultra-light materials for cars and funding for NSF to support research in advanced manufacturing areas such as nano-manufacturing, next-generation robotics and "smart" buildings and bridges.

- *The National Nanotechnology Initiative (NNI).* The NNI is the U.S. Federal government's interagency program for coordinating R&D and enhancing communication and collaborative activities in nanoscale science, engineering and technology.
- *National Digital Engineering and Manufacturing Consortium (NDEMC).* NDEMC is a public-private partnership launched in March 2011 that brings together manufacturers, industry associations, Federal agencies, national labs, and research universities to make modeling and simulation capabilities available to small-and medium-sized manufacturers.

The manufacturing sector would also greatly benefit from some of the policies outlined elsewhere in this report, such as robust basic research funding, an expanded and enhanced corporate R&D tax credit, and accelerated R&D, specifically in biotechnology, nanotechnology, clean energy and advanced manufacturing (Chapter 3); initiatives to support STEM education, such as the Skills for America's Future Initiative and the Department of Education's "Race to the Top Initiative" (Chapter 4); infrastructure investments (Chapters 5 and 7); and supporting Regional Innovation Clusters, the National Export Initiative, corporate tax reform, and an effective intellectual property regime (domestically and abroad) (Chapter 7).

References

Autor, David, David Dorn, and Gordon Hansen. 2011. "The China Syndrome: The Local Labor Market Effects of Import Competition" MIT Working Paper; econ-www.mit.edu/files/6613.

Bernard, Andrew B., J. Bradford Jensen and Peter K. Schott. 2004. "Facing the Dragon: Prospects for U.S. Manufacturers in the Coming Decade." Yale School of Management Working Paper, faculty.som.yale.edu /peterschott/ files /research/papers/dragon.pdf.

Board of Governors of the Federal Reserve System, (Accessed December 2011) Industrial Production and Capacity Utilization-G17, Industrial Production: Market, Industry Groups, and Individual Series, Data from January 1986 to present (Tables 1, 2, and 10); www.federalreserve.gov /releases/g17/table1_2.htm.

Bureau of Labor Statistics. (Accessed November 2011) Table B–1. Employees on nonfarm payrolls by major industry sector, 1961 to date; ftp://ftp.bls.gov/pub/suppl/empsit.ceseeb1.txt.

Bureau of Labor Statistics. (Accessed December 2011) Labor Productivity and Costs Data, Industry Labor Productivity and Costs: Indexes; www.bls.gov /lpc/iprprodydata.htm.

Bureau of Economic Analysis. 2011. *Survey of Current Business 2006–2009.* January; www.bea.gov/scb/pdf/ 2011/01January/0111_indy_accts_ tables. pdf.

Bureau of Economic Analysis. (Accessed December 2011). GDP-by-Industry Data. www.bea.gov/industry/ gdpbyind_data.htm.

Bureau of Economic Analysis. (Accessed December 2011). International Economic Accounts, Trade in Goods and Services 1992–Present; www.bea.gov/international/index.htm.

Council of Economic Advisers. 2011. *Economic Report of the President.* Washington, DC: U.S. Government Printing Office. February. www.gpoaccess.gov/eop/2011/pdf/ERP-2011.pdf.

Executive Office of the President of the United States. 2009. *A Framework for Revitalizing American Manufacturing.* December; www.whitehouse.gov /sites/default/files/microsites/20091216-manufacturing-framework.pdf.

Houseman, Susan, Christopher Kurz, Paul Lengermann and Benjamin Mandel. 2011. "Offshoring Bias in U.S. Manufacturing," *Journal of Economic Perspectives.* Vol. 25, no. 2. Spring; pubs.aeaweb.org/doi/pdfplus /10. 1257/ jep. 25.2.111.

Lacey, T. Alan, and Benjamin Wright. 2009. "Occupational Employment Projections to 2018." *Monthly Labor Review* (November): 82–123, Table 3.

National Center for Education Statistics. 2010. *2010 Digest of Education Statistics;* nces.ed.gov/programs/digest/ 2010menu_tables.asp.

National Center for Education Statistics. Integrated Postsecondary Education Data System (IPEDS). Spring 2006 to Spring 2010; nces.ed.gov /programs/digest/d10/tables/dt10_362.asp.

National Institutes of Standards and Technology. 2011. The Manufacturing Extension Partnership: Partnering for Manufacturing Innovation and Growth 2011, www.nist.gov/mep/upload/MEP-PARTNERING-IM PACTSFEB2011.pdf.

National Science Foundation, National Center for Science and Engineering Statistics. 2011. *Research and Development in Industry: 2006–07.*

Detailed Statistical Tables NSF 11–301. Arlington, VA. Available at www.nsf.gov/ statistics/nsf11301/.

President's Council of Advisors on Science and Technology. 2011. "Report to the President on Ensuring American Leadership in Advanced Manufacturing." June; www.whitehouse.gov/sites/default/files/microsites /ostp/pcastadvanced-manufacturing-june2011.pdf.

Tassey, Gregory. 2010. "Rationales and Mechanisms for Revitalizing U.S. Manufacturing R&D Strategies." *Journal of Technology Transfer 35*: 283–333.

Wolfe, Raymond M. 2009. "U.S. Business R&D Expenditures Increase in 2007; Small Companies Perform 19% of Nation's Business R&D." *InfoBrief NSF 09–316*. Washington, DC: National Science Foundation; www.nsf.gov/ statistics/infbrief/nsf09316/.

THE PRIVATE SECTOR AS THE ENGINE OF INNOVATION

Introduction

Increasing the competitiveness and the capacity of the United States to innovate goes beyond improving research, education, infrastructure, and the manufacturing sector. Many other factors can also lead to success, but perhaps chief among them is ensuring that both established firms and entrepreneurs in the private sector have the best possible environment in which to innovate. One of the major strengths of the American economy is that its decentralized, competitive markets provide the best method for determining the value of innovative opportunities and enabling their diffusion throughout the economy. This chapter explores the following areas that help provide a good environment for private sector innovation:[1]

- Support regional clusters
- Accelerate high-growth entrepreneurship through Startup America's public and private-sector initiatives
- Promote exports and access to foreign markets
- Restructure corporate taxes
- Provide an effective intellectual property system

Regional Clusters and Entrepreneurship

Despite more open markets, faster and cheaper transportation, and an increasingly robust digital infrastructure, location has continued to be central to competitiveness and innovation.[2] The prime example of how location still matters is regional clusters, which "are geographic concentrations of interconnected businesses, suppliers, service providers, coordinating intermediaries, and associated institutions like universities or community colleges in a particular field (e.g., information technology in Seattle, aircraft in Wichita, and advanced materials in Northeast Ohio)."[3] Regional clusters can also be thought of as an "innovation ecosystem" that "is made up of communities of people with different types of expertise and skill sets. Scientists, administrators, business leaders, engineers, writers, educators, health-care professionals, and other individuals all play a role."[4]

Once a critical mass of firms shares a common location they derive many types of advantages from this proximity. These firms develop close relationships, giving them better access to information and allowing them to interact more efficiently. Firms in clusters can more easily find workers with relevant experience, and suppliers may also cluster nearby, lowering input costs. These advantages are difficult, if not impossible, to take advantage of when firms are not close to each other.[5] Evidence indicates that areas with strong clusters perform better economically than areas without these clusters; they have higher job growth, higher wage growth, more businesses and a higher rate of patenting.[6]

Multiple studies highlight the positive correlation between the existence of regional innovation clusters (RICs) and wages. Wheaton and Lewis (2002) examined the effects of industrial and occupational specialization on manufacturing wage levels across 220 metropolitan areas.[7] They found that for the typical metropolitan area, a doubling in employment concentration in a particular industry (similar to what would occur when industry clusters are developed) is associated with a 2 percent increase in wages. Gibbs and Bernat (2001) found that wages for workers in industry clusters were about 6 percent higher than for workers in the same industry in a non-clustered location.[8] A Kansas cluster focused on aviation manufacturing and development provides yet another example of the wage benefits of RICs. The Kansas aviation cluster boasts 17.8 percent of all Kansas manufacturing employment, with average annual wages of $63,000, compared with $40,000 in average annual wages for all U.S. industries in 2006.[9]

RICs can also improve the productivity of firms operating in the clusters. Green-stone and Moretti (2004) evaluated the impact of large plants clustering together.[10] By comparing the productivity and employment growth of sites selected by large plants to those of "runner up" sites with similar characteristics, they found that firms clustering together increased productivity by 12 percent and employment by 9 percent.

New businesses are also generated by RICs; from the more than 150 clusters that exist around the country, RICs have resulted in increased spin-offs, creating new commercial activity. For example, the *CleanTECH San Diego* cluster initiative that was launched in 2007 and focused on energy efficiency, renewable energy, transportation, and water management has generated tremendous startup activity.

San Diego now boasts more than 800 clean technology companies, supported by world-class universities and a network of investors.

The clear economic benefits generated by RICs suggest the need to encourage the growth of these clusters. The Federal government is working in partnership with state and local efforts through agencies such as the Small Business Administration (SBA) and the U.S. Department of Commerce's Economic Development Administration (EDA). One example of EDA funding that helps clusters achieve superior results is a public-private partnership that led to the creation of a new proof-of-concept center at the University City Science Center in Philadelphia (see box 7.1). This example highlights innovation occurring at a regional, economic development level. Another example of an economic development agency working at the grass-roots level is NorTech based in Northern Ohio (see box 7.2).

Another Department of Commerce effort to promote entrepreneurship at the regional level is the establishment of the Office of Innovation and Entrepreneurship (OIE). The goal of the OIE is to promote innovation-based, high-growth entrepreneurship by increasing the efficiency and effectiveness of efforts to commercialize technology developed through university and federally funded research. The OIE manages the i6 Challenge, a multiagency competitive grant program that encourages innovative partnership models that accelerate technology commercialization, new venture formation, and job creation. It also manages the National Advisory Council on Innovation and Entrepreneurship, whose mission is to advise on the best methods to foster entrepreneurship and to develop innovation ecosystems such as RICs.

Other efforts include: the SBA's effort with the Department of Defense to develop clusters focusing on advanced technologies such as robotics, energy, and cyber-security; EDA's RIC efforts in areas such as best practices and 21st

century infrastructure, as well as its work through the Taskforce for the Advancement of Regional Innovation Clusters; and the Department of Agriculture's initiatives to bring regional strategies to rural areas that involve regional food systems, renewable energy, broadband, and recreation. Finally, another recent significant development is the reauthorization, for another 6 years, of the SBA's Small Business Innovation Research and Small Business Technology Transfer programs, which are set-aside programs for small businesses to engage in Federal R&D and to facilitate cooperative R&D between small businesses and research institutions, respectively. The programs were also expanded to allow firms that are majority owned by venture capital and that have private capital support to participate.

Box 7.1. Example of a Public-Private Partnership: QED

QED is a unique multi-institutional proof-of-concept mechanism that supports academic life science researchers as they transition their discoveries into products for end users. The University City Science Center in Philadelphia, Pennsylvania, (the oldest and largest urban research park in the United States) created the QED program in 2009 to bridge the gap between academic research grants and commercial seed funding. The goals of the program are to engage Greater Philadelphia's academic institutions, research scientists, entrepreneurs, investors, and industry in early-stage commercialization, and ultimately to increase the pace and value of technology transfer in the region by developing a pipeline of new technologies that could significantly improve human health.

QED provides key resources, including business guidance, bridge funding, and access to industry and investor representatives, to competitively selected projects. Currently, 19 research institutions participate in the program under a common set of terms and conditions that govern funding, indirect costs, intellectual property, and revenue sharing for program sustainability. Funding decisions are made by a regional selection team composed of representatives from pharmaceutical, medical device and medical diagnostics companies, private equity and venture capital firms, and economic development organizations. Each project selected for funding receives up to $200,000 over 12 months, with half of the funding provided by the Science Center and the other half by the scientist's host institution.

To date, QED has received and evaluated more than 227 proposals. Proof-ofconcept plans have been developed, with the assistance of business advisors, for 40 life science technologies at 15 institution, and 12 projects at eight institutions have been selected to receive funding. Of the nine projects that have been substantially completed, five have resulted in the licensing or optioning of technologies to the private sector, either through start-up or established companies. One of the licensed technologies represents the first example of technology from The Children's Hospital of Philadelphia, the Nation's first hospital for children, being commercialized via start-up company formation.

Currently in the fourth cycle of its pilot phase, QED has received funding from the Commonwealth of Pennsylvania's Ben Franklin Technology Development Authority, the William Penn Foundation, the U.S. Department of Commerce's Economic Development Administration and Wexford Science + Technology.

QED's early successes demonstrate the program's potential for meaningful impact on the region's innovation ecosystem through the collective engagement of academic, private sector, and entrepreneurial stakeholders. The program is both scalable and transferrable, and could serve as a template for similar efforts in other sectors and in other regions.

Box 7.2. Regional Innovation Clusters: NorTech

One excellent example of how regional innovation clusters can work to improve the economic climate of an area through the support of emerging technology industries is NorTech.1 A regional nonprofit technology-based economic development organization serving 21 counties in Northeast Ohio, NorTech is funded by public and private partners of regional businesses and philanthropic communities and supported by the U.S. Department of Commerce's Economic and Development Administration. As a catalyst for developing regional innovation clusters, NorTech is currently focused on two industries: advanced energy and flexible electronics.

The organization's cluster development model serves all organizations in the value chain and those that support the value chain—companies of all sizes; research institutions; universities; public, private and philanthropic funding sources; all levels of government, industry associations; and other economic development organizations.

NorTech's approach is to engage in activities at three levels: the cluster company and project level, the regional level, and the national level. Based on the Northeast Ohio's unique strengths and assets, NorTech drives the development of regional innovation clusters by:

- Attracting new members to the cluster by promoting Northeast Ohio's technology story;
- Building relationships among cluster members for funding, research, and revenue opportunities;
- Engaging with Federal and state governments and policy leaders to develop strategies to improve the likelihood clusters will continue to grow; and
- Collecting, reporting, and utilizing data that creates and influences cluster growth.

NorTech believes in the value of regional innovation clusters as a "bottoms up" approach to creating jobs and making the United States more globally competitive, specifically in Northeast Ohio. Clusters result in numerous benefits for a region such as creating new, higher wage jobs; providing regional business opportunities that are less susceptible to off-shoring; stabilizing diverse communities by repurposing idle assets and human capital; and increasing the export of regionally produced manufactured goods to other markets.

[1] See www.nortech.org/clusters/regional-innovation-cluster and www.nortech.org /about-us /what-we do for more information.

Startup America

In the United States, firms younger than 5 years create a significant fraction of new jobs.[11] However, many young firms struggle to survive beyond the startup period. The rate of new business startups has been declining over the past twoand-a-half decades, meaning fewer would-be entrepreneurs are rising to the challenge of turning new ideas into new businesses (see figure 7.1).

Launched in January 2011, Startup America is a White House initiative to accelerate high-growth entrepreneurship throughout the Nation. President

Obama has called on both the Federal government and the private sector to dramatically increase the prevalence and success of entrepreneurs across the country.

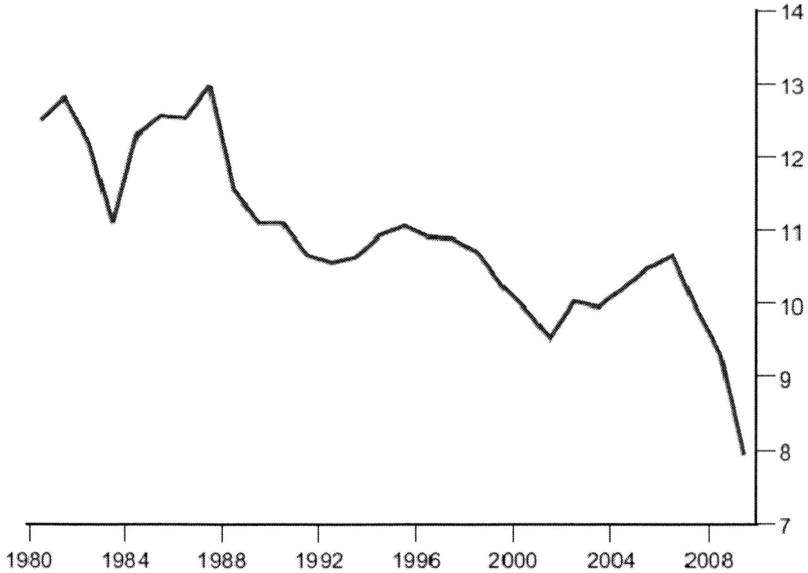

Source: U.S. Census Bureau, Center for Economic Studies, Business Dynamics Statistics.

Figure 7.1. U.S. Private Business Startup Rate, 1980–2009.

First, the Obama Administration's Startup America initiative is an ongoing series of policy actions to improve the environment for high-growth entrepreneurship in five key areas:

1) **Increasing access to capital** for high-growth companies (including zero capital gains tax on qualified small business investments and streamlined rules for private funds that invest in lower income communities);
2) **Creating mentorship and educational opportunities** for entrepreneurs (including new opportunities for clean energy entrepreneurs, military veterans, and undergraduate engineers);

3) **Reducing barriers** that can limit the growth of entrepreneurs through the solicitation of recommendations regarding the modification or elimination of regulations;
4) **Accelerating innovation** from lab to market for federally-funded R&D (including lower cost access to government-patented energy technology, and new funding for regional proof-of-concept centers and regionally interconnected networks of researchers, managers and capital across the business, education and government sectors); and
5) **Driving a nationwide effort** by the Administration to engage potential new opportunities in industries like healthcare, clean energy, and learning technologies.

Second, the Startup America Partnership has been launched, which consists of alliances of entrepreneurs, corporations, universities, foundations, and other leaders whose goal is to encourage innovative, high-growth U.S. startups. The Startup America Partnership has created a national online network where high-growth entrepreneurs can establish free membership profiles and unlock resources from dozens of companies—from free software to free business filing to steeply discounted computer hardware. The total value of these resources is over $730 million and climbing.

Promoting America's Exports and Improving Access to Foreign Markets

A vibrant and expanding export market is essential for economic growth and for creating new jobs. Many of the recommendations previously discussed to increase innovation in the United States may also lead to more exports, which in turn should stimulate further innovation in the United States. If the United States can produce higher quality goods and services more efficiently, it will be more competitive in foreign markets. However, U.S. exporters, particularly manufacturing firms, often are not only competing against private sector domestic firms but are also competing against foreign firms that may benefit from foreign government support for particular manufacturing sectors. For example, in some instances, countries do not allow the foreign exchange rates of their currencies to be fully flexible and market determined. This can make U.S. goods more expensive than they otherwise would be, limiting U.S. export growth.[12]

Ensuring that U.S. businesses have fair and open access to foreign markets is an important component of increasing U.S. exports. Enforcing the obligations of other countries with respect to market access cannot fall to businesses that export but must be done by the U.S. government. This is yet another area where there is a clear role for government to improve the competitiveness of the United States. Some progress already has been made on this front.

Therefore, in addition to pursuing policies to improve innovation, the Federal government can play a role in promoting U.S. exports. In March 2010, President Obama launched the National Export Initiative (NEI), which "brings a sustained, vigorous commitment to ensure fair and open export market for American businesses" and is "an ambitious effort to help American businesses that sell their goods and services abroad. By unlocking foreign markets for U.S. goods and services, improving access to credit for U.S. businesses, and undertaking other measures, the NEI seeks to double U.S. exports in five years and support millions of additional jobs."[13]

Additionally, Congress approved three free trade agreements, with Panama, Colombia, and South Korea in quick succession in the fall of 2011, marking the biggest step forward in opening foreign markets to American goods and services since the North American Free Trade Agreement and the Uruguay Round of the mid-1990s. Of these agreements, the most commercially significant was the Korea-United States free trade agreement (KORUS). A study by the International Trade Commission estimated that the renegotiated agreement with Korea could boost annual U.S. goods exports to Korea by as much as $11 billion.[14] The agreement also included Korean commitments expected to result in considerable expansion of U.S. services exports.

In November 2009, President Obama announced the United States' intention to participate in the Trans-Pacific Partnership (TPP) negotiations, with the goal of concluding a high-standard free trade agreement with countries in the huge and growing markets of the Asia-Pacific region. This next-generation agreement would address not only the core issues traditionally included in trade agreements, but also new issues such as making the regulatory systems of TPP countries more compatible so U.S. companies can operate more seamlessly in TPP markets, and helping innovative, job-creating small and medium-sized enterprises participate more actively in international trade and in investment in innovative products and services, including digital technologies, and mechanisms to ensure state-owned enterprises compete fairly with private companies. In addition to the United States, the other countries participating in the negotiations currently include Australia, Chile, Peru, Singapore, Brunei Darussalam, Malaysia, New Zealand, and Vietnam. Ten rounds of negotiations among these prospective partners have already taken place, with the most recent round having been held in Malaysia in December 2011. In November 2011, Japan, Canada, and Mexico announced their interest in joining the negotiations.

The costs of financing export operations pose an additional barrier for smaller firms. Financial institutions may erroneously regard a small firm that is highly dependent on exports as a riskier borrower than one that is entirely domestic in its focus. The mission of the Export-Import Bank (Ex-Im), along with other institutions, is to proactively support small and medium-sized firms. In fiscal year 2010, Ex-Im authorized $5 billion—20 percent of authorizations—to support small businesses as primary exporters.[15] The two Ex-Im products most used by U.S. small businesses are export-credit insurance and working-capital guarantees. Export-credit insurance protects exporters and lenders from the risk of buyer nonpayment for commercial or political reasons and enables exporters to extend credit to international customers. Working-capital guarantees cover 90 percent of the outstanding balance of working-capital loans to exporters supported by export-related inventory and accounts receivable. In fiscal year 2010, the Bank issued 2,524 insurance policies to small-business exporters—90 percent of the total number of policies for the year. The Bank also authorized a record $2.2 billion in working-capital guarantees, 70 percent of which supported small business.[16]

Corporate Taxes

The United States has the second-highest statutory corporate income tax rate in the Organisation for Economic Co-operation and Development (OECD). Japan has the highest. However, the United States does not rank as high in terms of the average effective tax rate paid by corporations.[17] One reason is that the corporate tax code has numerous provisions for special deductions, credits, and other tax expenditures that benefit certain activities.

The combination of a high statutory rate and numerous deductions and exclusions results in an inefficient tax system. The high statutory rate discourages saving and investment, while the features that limit the tax base favor debt over equity, encourage investment in certain favored assets over other kinds of investment, and drive capital out of the corporate sector into noncorporate forms of business. There are also inefficiencies due to the way the United States taxes the foreign income of U.S. multinational corporations. The lower foreign corporate tax rates, along with the fact that other countries use a territorial system of corporate taxation, places U.S. multinational companies at a cost disadvantage.

Finally, according to the President's Economic Recovery Advisory Board, the complexity of the code and its incentives for tax avoidance result in costs to firms that are "estimated to exceed $40 billion per year or more than 12 percent of the revenues collected. All of these factors act to reduce the productivity of American businesses and American workers, increase the likelihood and cost of financial distress, and drain resources away from more valuable uses."[18]

Given the inefficiencies described above, proposals to reform the corporate tax code would likely trade a lower statutory rate for a broader tax (that is, fewer provisions that favor one type of investment over another) while also, perhaps, dealing with the unequal treatment of U.S. multinationals relative to other countries. However, there are tradeoffs to moving to a more simplified corporate tax code, and changes could dampen innovation.[19] For example, R&D currently receives preferential treatment through a tax credit, and the Administration has argued for simplifying, enhancing, and expanding the R&D tax credit as a way of helping companies create jobs and increase productivity.[20]

Ensuring a Well-Functioning Intellectual Property Rights System

A well-functioning intellectual property rights (IPR) system is crucial for encouraging innovation and creating jobs. "Absent effective legal protections for innovators, other businesses can immediately exploit an innovator's idea, undermining the incentive to invent in the first place. Public policy solves this problem through intellectual property rights—allowing limited, short-run grants of exclusive rights to catalyze inventive activity."[21] And to safeguard those intellectual property rights, the Administration issued a White Paper in March 2011 with 20 recommendations for legislative changes based on its comprehensive review of existing law in order to ensure that American workers and businesses are protected, exemplifying the Administration's commitment to grow jobs and exports as well as to protect the health and safety of the American people.[22]

In the United States, intellectual property (IP) significantly influences innovation and economic growth. Industries that are the most intensive users of IP protections directly support millions of jobs across all sectors of the economy. Unfortunately, the U.S. patent system has not always functioned in a manner conducive to encouraging innovation.[23] In particular, it is crucial that the United States improve its IP system by reducing both review times as well as the cost of litigation related to patents. Fortunately, significant progress has been made in reforming the patent system in the United States. With the passage of the *America Invents Act* in September 2011, the United States Patent and Trademark Office (USPTO) will be able to offer, under a prioritized examination process, a new fast track for reviewing patents with a guaranteed 12-month approval timetable for certain patents.[24] Additional resources are provided in the Act, allowing USPTO to continue reducing the backlog of patent applications and the time it takes to review them. USPTO will offer entrepreneurs new ways to make litigation regarding patent validity less burdensome and at costs significantly less expensive than going to court.

IP protection abroad is also crucial for U.S. firms. Infringement of IPR in markets abroad causes significant financial losses for rights holders and legitimate businesses around the world and undermines key U.S. comparative advantages in innovation and creativity to the detriment of American businesses and workers. The Administration's *Joint Strategic Plan on Intellectual Property Enforcement*, issued in June 2010 by the White House Intellectual Property Enforcement Coordinator, lays out a comprehensive strategy for the U.S. Government to strengthen enforcement of intellectual property rights, both at home and abroad.[25] Initiatives on the international front include the U.S. government aggressively pursuing meaningful improvements in the protection and enforcement of U.S. intellectual property with our trading partners. This includes direct bilateral engagement to increase enforcement, participation in regional and multilateral fora, and the negotiation of new IPRs related instruments, such as the Anti-Counterfeiting Trade Agreement, and, where appropriate, enforcing our rights using the dispute settlement procedures of the World Trade Organization.

The U.S. government is also alert to emerging concerns regarding innovation and industrial policies in some of our trading partners that may disadvantage U.S. IP rights holders. Such policies include measures that condition government benefits on the local development or ownership of IPR, or that condition market access or other benefits on the transfer of technology, IPR or other proprietary information from foreign companies to domestic entities. They may also include measures to restrict the ability of U.S. rights holders to freely negotiate the terms and conditions of the use of their IPR or impediments to enforce contractual ar rangements.

The Obama Administration is committed to an intellectual property rights system that recognizes that IP rights are fully consistent with—and indeed enable—other core values such as the norms of legitimate competition, free speech, fair process, and the privacy of users. The Administration is also committed to addressing international health and public safety challenges. For example, the USPTO has issued a request for information to develop strategies to incentivize humanitarian technologies through the intellectual property system.

REFERENCES

Atkinson, Robert D. 2011. *The Case for Incentives in the U.S. Corporate Tax Code. Washington*, DC: The Information Technology and Innovation Foundation, September 27; www.itif.org/files/2011-atkinson-tax-incentives.pdf.

Center for Economic Development and Business Research. 2008. *Kansas Aviation Manufacturing*. Wichita, KS: W. Frank Barton School of Business, Wichita State University, September; webfiles.wichita.edu /cedbr/AvFINAL.pdf.

Delgado, Mercedes, Michael E. Porter, and Scott Stern. 2011., *Clusters, Convergence, and Economic Performance*. (March); www.isc.hbs.edu/pdf /DPS_Clusters_Performance_2011-0311.pdf. [See reference in chapter 1.]

Estrin, Judy. 2009. *Closing the Innovation Gap: Reigniting the Spark of Creativity in a Global Economy*. New York: McGraw Hill.

The Export Promotion Cabinet. 2010. *Report to the President on the National Export Initiative: The Export Promotion Cabinet's Plan for Doubling U.S. Exports in Five Years*. Washington, DC: U.S. Department of Commerce, September 2010; www.whitehouse.gov/sites/default/files/nei_report_9-16-10_full.pdf.

Ezell, Stephen J., and Robert D. Atkinson. 2011. *The Case for a National Manufacturing Strategy*. Washington, DC: The Information Technology and Innovation Foundation, April.

Gibbs, Robert M., and G. Andrew Bernat, Jr. 2001. "Rural Industry Clusters Raise Local Earnings." *Rural Development Perspectives* 12 (March): 18–25; www.ers.usda.gov/Publications/RDP/RDP697/RDP697d.pdf.

Greenstone, Michael, and Enrico Moretti. 2004. "Bidding for Industrial Plants: Does Winning a 'Million Dollar Plant' Increase Welfare?" Department of Economics Working Paper 04–39. Cambridge, MA: Massachusetts Institute of Technology.

Greenstone, Michael, Richard Hornbeck and Enrico Moretti, "Identifying Agglomeration Spillovers: Evidence from Winners and Losers of Large Plant Openings" *Journal of Political Economy*, 118(3), 2010.

Haltiwanger, John C., Jarmin, Ron S. and Miranda, Javier. 2010. *Who Creates Jobs? Small vs. Large vs. Young. National Bureau of Economic Research,* Working Paper No. 16300 (August 2010, Revised November 2011): 24–25; www.nber.org/papers/w16300.pdf.

Muro, Mark, and Bruce Katz. 2010. "The New 'Cluster Moment': How Regional Innovation Clusters Can Foster the Next Economy." Metropolitan Policy Program at Brookings Institution, September 2010. www.brookings.edu/~/media/Files/rc/papers/2010/0921_clusters_muro_katz/0921_clusters_muro_katz.pdf.

National Academy of Sciences, National Academy of Engineering, and Institute of Medicine. 2010. *Rising Above the Gathering Storm, Revisited: Rapidly Approaching Category 5.* Washington, DC: National Academies Press.

National Economic Council, Council of Economic Advisers, and Office of Science and Technology Policy. 2011. *A Strategy for American Innovation: Securing Our Economic Growth and Prosperity. Washington, DC*, February.

Porter, Michael E. 1998. "Clusters and the New Economics of Competition." *Harvard Business Review* www.nap. edu/catalog/11463.html(November–December): 81–83.

The President's Economic Recovery Advisory Board. 2010. *The Report on Tax Reform Options: Simplification, Compliance, and Corporate Taxation.* Washington, DC: White House, August; www.whitehouse.gov/sites/ default/files/microsites/PERAB_Tax_Reform_Report.pdf.

United States International Trade Commission (USITC). *U.S.-Korea Free Trade Agreement: Potential Economy-wide and Selected Sectoral Effects.* Investigation No. TA-2104-24. USITC Publication 3949. September 2007; www.usitc.gov/publications/docs/pubs/2104F/pub3949.pdf.

Wheaton, William C., and Mark J. Lewis. 2002. "Urban Wages and Labor Market Agglomeration." *Journal of Urban Economics* 51 (May): 542–562.

White House. Office of the Press Secretary. 2010. *Joint Strategic Plan on Intellectual Property Enforcement.* 2010; www.whitehouse.gov/sites /default/files/omb/assets/intellectualproperty/intellectualproperty_strategic _plan. pdf.

White House. Office of the Press Secretary. 2011a. "Administration's White Paper on Intellectual Property Enforcement Legislative Recommen dations." March 2011; www.whitehouse.gov/sites/default/files /ip_white_ paper.pdf.

White House. Office of the Press Secretary. 2011b. "President Obama Signs America Invents Act, Overhauling the Patent System to Stimulate Economic Growth, and Announces New Steps to Help Entrepreneurs Create Jobs." Press Release, September 16; www.whitehouse.gov/the-press-office/2011/09/16/president-obama-signsamerica-invents-act-overhauling-patent-system-stim.

MOVING FORWARD

Throughout its history, the United States has faced numerous challenges that have threatened to derail its economic growth and prosperity. However, the United States always has been able to meet and overcome these challenges, and in so doing, increase the standard of living of its citizens. The private sector has been the primary driver of this increased prosperity, as businesses innovate to remain competitive.

Working with the private sector, government has also played a key role in supporting innovation by providing the necessary building blocks. In particular, the Federal government has provided funding and support for basic research, funding that has been important to many of the major innovations of the 20th century.

The Federal government also helped encourage the creation of arguably the world's leading system of higher education. First-rate colleges and universities train the workers needed to lead innovative activities in the private sector. The infrastructure needed by business to innovate and compete, from railroads in the 19th century to broadband Internet networks in the late 20th and early 21st century, was built with support from the Federal government. In these three areas a government role is necessary, as the private sector will not invest sufficiently on its own.

In the first decade of the 21st century, the U.S. economy was no longer growing as rapidly as it had in the past: job creation slowed, and income levels stagnated for large segments of the population. It is no coincidence that the ability of the United States to innovate also suffered during this period. Federal support for basic research has not kept pace with the growth of the economy, the education system has not done a good enough job preparing students to become skilled workers, and the nation's infrastructure has not kept up with growing needs of the U.S. population and U.S. businesses.

Other factors have also diminished the innovative capacity of the United States. The manufacturing sector, a key driver of innovation in the past, has been experiencing a long period of decline.

At the same time, the United States has had difficulty accessing certain foreign markets, enforcing intellectual property rights around the world, and achieving a balanced tax system. Each of these factors, as well as others highlighted in this report, need to be addressed if the United States is to regain its preeminent innovative capacity.

Although the list of problems is long, and though it will take time to overcome them, there is also a long list of policy tools that will allow the United States to address and correct these problems.

Any sensible and successful approach to overcoming these problems must start by implementing the following 10 key policy proposals:

1. Continue to Support Government Funding for Basic Research

For the United States to maintain a leadership role in innovation, it is critically important that the Federal government continue its support for basic research. Also, since quality scientific education and scientific advances take many years, investments in research should be stable to improve career prospects of new science doctorates and to encourage younger students to choose science as a career.

2. Enhance and Extend the R&D Tax Credit

Although the Federal government's role in R&D is crucial, private R&D investment remains important and a simplified, enhanced, and extended corporate R&D tax credit would create the proper incentives for private industry to undertake the risks associated with R&D spending.

3. Speed the Movement of Ideas from Basic Science Labs to Commercial Application

Entrepreneurs can find it difficult to get early-stage funding for their ideas. Other barriers to commercialization exist, such as lack of business experience on the part of would-be entrepreneurs. "Proof of Concept" centers can help overcome this barrier by supporting entrepreneurs at all stages of the development process and these centers need further encouragement.

The Administration is committed to continuing its i6 Green Challenges to help develop these centers. Other initiatives that should be encouraged include the Advanced Manufacturing Partnership where industry, academia and government can collaborate and accelerate the development of emerging technologies.

4. Address STEM Shortcomings

Poor STEM participation and performance in the nation's schools must be remedied, as students are leaving secondary schools poorly trained to continue studying in STEM fields.

One avenue to address these problems is initiatives such as "Educate to Innovate," public-private partnerships that expand STEM education to all students, particularly those of underrepresented groups, through interactive games and other methods.

Another avenue to promote and prepare disadvantaged youth and dislocated workers for STEM careers, while simultaneously enhancing the competitive position of local and regional employers, was DOL's STEM Opportunities in the Workforce System Initiative.

These 2009 grants focused primarily on expanding and aligning current and new STEM workforce education and training strategies, activities, and resources in One-Stop Career Centers. Also, additional funding is needed to train more STEM teachers.

Programs such as NSF's Widening Implementation and Demonstration of Evidence based Reforms (WIDER) should be implemented to improve undergraduate STEM instruction and outcomes at universities.

5. Increase Spectrum for Wireless Communications

The United States faces a spectrum crunch in the coming years, which could severely constrain innovation. The goals set by the "National Wireless Initiative," include doubling the amount of spectrum available for wireless broadband services and helping rural areas gain access to wireless broadband services.

6. Increase Access to Data to Help Spur Innovation

Open access to data is a crucial component of a successful innovation policy, and steps taken to encourage this include the launch of data.gov, a platform that provides public access to valuable datasets; an initiative to simplify access to high value data by, for example, creating standards; and the use of challenges and prizes to bring together communities of innovators to help spur new technologies. These efforts need to be continued and expanded.

7. Coordinate Federal Support for Manufacturing

For the manufacturing sector to reverse its decline, it is vital to continue funding and supporting manufacturing specific programs like NIST's MEP, SelectUSA, and the individual pieces of the Advanced Manufacturing Partnership. In addition, it is important to re-focus and improve coordination of manufacturing programs under the Office of Manufacturing Policy's new structure led by co-chair's NEC Director Sperling and Commerce Secretary Bryson.

8. Continue and Strengthen Efforts to Foster Regional Clusters and Entrepreneurship

Evidence shows that regional innovation clusters increase jobs and wages. Multiple efforts are already under way within the Federal government to promote and encourage entrepreneurship and clusters and these efforts must continue. In the area of encouraging clusters, efforts include the i6 Challenge (a competitive grant program that encourages innovative partnership models), EDA's efforts through the Taskforce for the Advancement of Regional

Innovation Clusters, the Department of Agriculture's initiatives to bring regional strategies to rural areas and the recently reauthorized SBA Small Business Innovation Research and Small Business Technology Transfer programs. To encourage entrepreneurs, the Startup America initiative is increasing access to capital and facilitating mentorships and the Startup America Partnership has launched an online network that provides entrepreneurs access to valuable resources from dozens of companies. Efforts like these will need continued support in the years ahead in order to ensure entrepreneurs have the resources they need to help drive innovation.

9. Promote America's Exports and Improve access to Foreign Markets

It is vital that U.S. businesses have fair and open access to foreign markets. To help ensure firms have this access, the Administration launched the National Export Initiative (NEI), and Congress enacted legislation the President submitted to implement free trade agreements with Panama, Colombia, and South Korea. To build on this momentum, the United States is participating in the Trans-Pacific Partnership negotiations, a free trade agreement with key partners in the Asia-Pacific region. This agreement, when finalized, will be a significant step forward as it not only addresses traditional trade issues, but also includes regulatory harmonization, trade and investment in innovative products and services (including digital technologies), and mechanisms to ensure state-owned enterprises compete fairly with private companies.

10. Ensure That the Conditions Exist in Which Private Enterprise Can Thrive

The private sector is the engine of innovation in the United States and it is crucial that both established firms and entrepreneurs in the private sector have the best possible environment in which to innovate. To this end, areas that should be the focus of attention in the United States in the years ahead include reforming the corporate tax system, and ensuring that the intellectual property system continues to function in a way that encourages growth.

The United States is facing economic challenges as important and concerning as any we have faced in our history. Meeting these challenges will require effort and the enactment of policies, such as those listed above and others mentioned throughout this report. However, there is little doubt that the United States can meet these challenges and subsequently become more innovative and competitive, providing new jobs, new businesses, and new industries.

SUPPLEMENTAL MATERIALS

Innovation Advisory Board Members

Robert Atkinson
Founder and President,
Information Technology and Innovation Foundation

Rebecca O. Bagley
President & CEO, NorTech

Jim Clements
President, West Virginia University

Abby Joseph Cohen
President of the Global Markets Institute and
Senior Investment Strategist,
Goldman Sachs

Larry Cohen
President, Communications Workers of America

Judy Estrin
CEO, JLabs

Rebecca Henderson
Senator John Heinz Professor of Environmental Management
at the Harvard Business School

Irwin Jacobs
Director (Cofounder and former CEO), Qualcomm

Arthur Levinson
Chairman, Genentech, Inc.; Apple, Inc.

James Manyika
Director (Senior Partner), McKinsey & Company; Director, McKinsey Global Institute

Natalia Olson-Urtecho
President & CEO, EG (formerly Ecolibrium Group)

Kim Polese
Chairman, ClearStreet Inc.

Lucy Sanders
CEO and Co-Founder, National Center for Women & Information Technology

Julie Shimer
President & CEO, Welch Allyn

Stephen Tang
President, CEO & Director, University City Science Center

Section 604 of the America COMPETES Reauthorization Act of 2010

Study on Economic Competitiveness and Innovative Capacity of United States and Development of National Economic Competitiveness Strategy

(a) Study-

(1) IN GENERAL- Not later than 1 year after the date of the enactment of this Act, the Secretary of Commerce shall complete a comprehensive study of the economic competitiveness and innovative capacity of the United States.

(2) MATTERS COVERED- The study required by paragraph (1) shall include the following:

(A) An analysis of the United States economy and innovation infrastructure.

(B) An assessment of the following:

(i) The current competitive and innovation performance of the United States economy relative to other countries that compete economically with the United States.

(ii) Economic competitiveness and domestic innovation in the current business climate, including tax and Federal regulatory policy.

(iii) The business climate of the United States and those of other countries that compete economically with the United States.

(iv) Regional issues that influence the economic competitiveness and innovation capacity of the United States, including—

(i) the roles of State and local governments and institutions of higher education; and

(ii) regional factors that contribute positively to innovation.

(v) The effectiveness of the Federal Government in supporting and promoting economic competitiveness and innovation, including any duplicative efforts of, or gaps in coverage between, Federal agencies and departments.

(vi) Barriers to competitiveness in newly emerging business or technology sectors, factors influencing underperforming economic sectors, unique issues facing small and medium enterprises, and barriers to the development and evolution of start-ups, firms, and industries.

(vii) The effects of domestic and international trade policy on the competitiveness of the United States and the United States economy.

(viii) United States export promotion and export finance programs relative to export promotion and export finance programs of other countries that compete economically with the United States, including Canada, France, Germany, Italy, Japan, Korea, and the United Kingdom, with noting of export promotion and export finance programs carried out by such countries that are not analogous to any programs carried out by the United States.

(ix) The effectiveness of current policies and programs affecting exports, including an assessment of Federal trade restrictions and State and Federal export promotion activities.

(x) The effectiveness of the Federal Government and Federally funded research and development centers in supporting and promoting technology commercialization and technology transfer.

(xi) Domestic and international intellectual property policies and practices.

(xii) Manufacturing capacity, logistics, and supply chain dynamics of major export sectors, including access to a skilled workforce, physical infrastructure, and broadband network infrastructure.

(xiii) Federal and State policies relating to science, technology, and education and other relevant Federal and State policies designed to promote commercial innovation, including immigration policies.

(C) Development of recommendations on the following:

(i) How the United States should invest in human capital.

(ii) How the United States should facilitate entrepreneurship and innovation.

(iii) How best to develop opportunities for locally and regionally driven innovation by providing Federal support.

(iv) How best to strengthen the economic infrastructure and industrial base of the United States.

(v) How to improve the international competitiveness of the United States.

(3) CONSULTATION-

(A) IN GENERAL- The study required by paragraph (1) shall be conducted in consultation with the National Economic Council of the Office of Policy Development, such Federal agencies as the Secretary considers appropriate, and the Innovation Advisory Board established under subparagraph (B). The Secretary shall also establish a process for obtaining comments from the public.

(B) INNOVATION ADVISORY BOARD-

(i) IN GENERAL- The Secretary shall establish an Innovation Advisory Board for purposes of obtaining advice with respect to the conduct of the study required by paragraph (1).

(ii) COMPOSITION- The Advisory Board established under clause (i) shall be comprised of 15 members, appointed by the Secretary—

(I) who shall represent all major industry sectors;

(II) a majority of whom should be from private industry, including large and small firms, representing advanced technology sectors and more traditional sectors that use technology; and

(III) who may include economic or innovation policy experts, State and local government officials active in technology-based economic development, and representatives from higher education.

(iii) EXEMPTION FROM FACA- The Federal Advisory Committee Act (5 U.S.C. App.) shall not apply to the advisory board established under clause (i).

(b) Strategy-

(1) IN GENERAL- Not later than 1 year after the completion of the study required by subsection (a), the Secretary shall develop, based on the study required by subsection (a)(1), a national 10-year strategy to strengthen the

innovative and competitive capacity of the Federal Government, State and local governments, United States institutions of higher education, and the private sector of the United States.

(2) ELEMENTS- The strategy required by paragraph (1) shall include the following:

(A) Actions to be taken by individual Federal agencies and departments to improve competitiveness.

(B) Proposed legislative actions for consideration by Congress.

(C) Annual goals and milestones for the 10-year period of the strategy.

(D) A plan for monitoring the progress of the Federal Government with respect to improving conditions for innovation and the competitiveness of the United States.

(c) Report-

(1) IN GENERAL- Upon the completion of the strategy required by subsection (b), the Secretary of Commerce shall submit to Congress and the President a report on the study conducted under subsection (a) and the strategy developed under subsection (b).

(2) ELEMENTS- The report required by paragraph (1) shall include the following:

(A) The findings of the Secretary with respect to the study conducted under subsection (a).

(B) The strategy required by subsection (b).

Endnotes for Rising to the Challenge

[1] International Monetary Fund, World Economic Outlook database, September 2011.

[2] U.S. Bureau of Economic Analysis 2010, Table 678. Note: In chained 2005 dollars.

[3] National Center for Health Statistics 2011, 134.

[4] Economic Research Service 2011.

[5] Nobelprize.org, *Nobel Prize Facts* www.nobelprize.org/nobel_prizes/nobelprize_facts.html.

[6] Thomson Reuters 2011.

[7] See, for example, National Academy of Sciences 2007.

[8] Charlton Research Company 2011.

[9] Peter D. Hart Research Associates, Inc. and The Winston Group 2006, 2.

[10] Manyika et al. 2011, 1.

[11] Based on research by Atkinson, Piketty, and Saez 2011 and Piketty and Saez 2003.

[12] Much has been written on the subject of the forces behind changes in wages and relative wages over the past several decades. Reasons offered for the various wage trends include the decline in the fraction of workers covered by collective bargaining, increased international competition, technological change, immigration, minimum wages, among others. See, for example, Sachdev 2007 for tables updated to 2008.

[13] U.S. Census Bureau, Foreign Trade Division, U.S. International Trade in Goods and Services (FT900) www. census.gov/foreign-trade.

[14] For the complete definition of this sector, see U.S. Census Bureau, Foreign Trade Division 2011.

[15] The manufacturing sector is not the only part of the economy that is exposed to international competition, as documented by Jensen (2011). The U.S. has run a trade surplus in services, and we export about $500 billion a year in services (compared with $565 billion in manufactured goods), and increasingly the service sector will be open to international competition which may create additional opportunities of U.S service exports. Jensen estimates that roughly a third of the service sector is subject to international competition.

[16] *The Atlantic Century* 2009.

[17] National Academy of Sciences 2007.

[18] National Academy of Sciences 2010.

[19] Fleischman, H.L., Hopstock, P.J., Pelczar, M.P., and Shelley, B.E. 2010.

[20] Delgado, Porter, and Stern 2011.

Endnotes for Keys to Innovation, Competitiveness, and Jobs

[1] The Advisory Committee on Measuring Innovation in the 21st Century Economy 2008, i.

[2] Historically, these two measures have been used as proxies for innovation, but recently efforts have been made to measure innovation more directly through innovation surveys. See www.nsf.gov/statistics/infbrief/ nsf09304/. It should also be mentioned that there are some objections to these proxies. For example, the OECD, in its guidelines on collecting and interpreting innovation data (often referred to as the "Oslo Manual") states that patents are not good proxies for innovation because they are inputs to innovation rather than outputs and because patents can lack any economic value. However, even the OECD recognizes that a deeper understanding of innovation necessarily requires learning more about patents.

[3] For an explanation of productivity change see Jorgenson and Griliches 1967, 249–283. For discussion of intangible capital and economic growth see Corrado, Hulten and Sichel 2009. See also Bureau of Labor Statistics multifactor productivity news releases 2011a, 2011b, and 2011c.

[4] Manyika et al. 2010, 10.

[5] World Economic Forum 2011–2012, 4. (WEF) quantifies a wide variety of factors under its "12 Pillars of Competitiveness." Those pillars are: (1) Institutions; (2) Infrastructure; (3) Macroeconomic environment; (4) Health and primary education; (5) Higher education and training; (6) Goods market efficiency; (7) Labor market efficiency; (8) Financial market development; (9) Technological readiness; (10) Market size; (11) Business sophistication; and (12) Innovation. According to the WEF Global Competitiveness Report 2011–2012, the United States ranked fourth overall in 2010 and then fifth in 2011. However, the factors that went into the WEF ranking, how those factors are computed, and then how the factors are added together all require subjective judgments.

[6] World Economic Forum 2011–2012, 8.

[7] Bernard, Jensen, and Schott 2009, 514.

[8] Cockburn, Stern, and Zausner 2011, 115.

[9] Murphy and Topel 2006.

[10] Committee on Science, Engineering, and Public Policy 1999, 31.

[11] Figure 1.4, Goldin and Katz 2008, 20.
[12] Goldin and Katz 2008, 19.
[13] Goldin and Katz 2008, 26.
[14] Figure 1.7, Goldin and Katz 2008, 24.
[15] Times Higher Education 2011–2012.
[16] OECD Indicators 2011, 321.
[17] U.S. Environmental Protection Agency 2000.
[18] Finn 2002.

End Notes for Federal Support for Research and Development

[1] These institutions make up what is sometimes called a country's "innovative capacity." See, for example, Furman, Porter, and Scott 2002.
[2] Atkinson and Andes 2009 cited in National Economic Council 2011, *A Strategy for American Innovation*, 8.
[3] The Advisory Committee on Measuring Innovation in the 21st Century Economy 2008.
[4] See Abramowitz 1956 and Solow 1957.
[5] Romer 1986 and 1990 is often credited with reviving interest in the study of inventiondriven growth. A lead ing textbook that reviews the nowextensive literature in this domain, both theoretical and empirical, is the work by Acemoglu 2008.
[6] Important papers by Arrow 1962 and Griliches 1957, 1958, 1979, among many others, recognized that techni cal progress was the outcome of deliberate investment in R&D or in refinement of a production process.
[7] Since the private sector along with privatepublic partnerships play an important role in innovation and R&D, the Administration is committed to providing the incentives for the private sector to engage in R&D to help fuel continued innovation in the United States. For example, the Administration is committed to extending the R&D tax credit.
[8] Muro and Katz 2010, 11.
[9] See endnote 30 in Muro and Katz 2010, 51.
[10] Delgado, Porter, and Stern 2011.
[11] Office of Science and Technology Policy 2011 and White House, Office of the Press Secretary 2011a.
[12] Burrows 1992 cited in Irwin and Klenow 1996, 12740.
[13] Regulatory decisions by the Federal Communications Commission, such as exempting early providers from paying access charges and allowing terminals to be connected to the telecommunications network, were also in strumental in promoting the development of the Internet.
[14] Nordhaus 2005.
[15] Council of 16 Economic Advisers 2011, 56.

[16] National Science Foundation 2010.

[17] For one independent study of this slowdown, see Litan, Mitchell, and Reedy 2007.

[18] For the definition of this sector, see U.S. Census Bureau 2010.

[19] Hockfield 2011.

[20] See U.S. Census Bureau for data on foreign trade of advanced technology products, www.census.gov/foreign trade/balance/c0007.html.

[21] National Economic Council, *A Strategy for American Innovation*, 8.

[22] World Economic Forum 2011, 514.

[23] National Economic Council, *A Strategy for American Innovation*, 8.

[24] Economic Development Administration Press Release 2011.

[25] White House, Office of the Press Secretary 2011b.

[26] For additional information, see National Nanotechnology Initiative, *What is Nanotechnology?*

[27] For additional information, see National Nanotechnology Initiative, *Benefits and Applications.*

[28] U.S. Department of Health and Human Services 2010.

[29] The Advisory Committee on Measuring Innovation in the 21st Century Economy 2008.

[30] The Advisory Committee on Measuring Innovation in the 21st Century Economy 2008.

[31] Abramowitz 1956 and Solow 1957.

[32] Romer 1986, 1990 and Acemoglu 2008.

[33] Arrow 1962 and Griliches 1957, 1958, 1979.

[34] Thomas Jefferson, in his letter to Isaac McPherson, dated 13 August 1813.

End Notes for Educating Our Workforce

[1] Becker 2008.

[2] Goldin and Katz 2008.

[3] These include skills such as mathematics, science, critical thinking, active learning, complex problem solving, operations analysis, systems analysis, and problem solving. See Carnevale, Smith, and Melton 2011, 7–10.

[4] Baumol, Schilling, and Wolff 2009, 723–724.

[5] Langdon et al. 2011, 3.

[6] Carnevale, Smith, and Melton 2011, 7–10 and 10–60.

[7] Council of Economic Advisers 2009, 9-10.

[8] Langdon et al. 2011, 4.

[9] National Science Foundation 2010, 3.17–3.20.

[10] Langdon et al. 2011, 4–5.

[11] Times Higher Education 2011.

[12] Institute of International Education 2011.

[13] National Center for Education Statistics 2010a, Table 241.

[14] Adelman 1998.

[15] Council of Economic Advisers 2010, 223–225.
[16] Organisation of Economic Co-operation and Development (OECD) 2011b.
[17] Carnevale, Smith, and Melton 2011, 46–47.
[18] Silverman and Light 2011. In addition, many research papers have found that the quality of undergraduate teaching is a factor that influences the number of STEM students. See for example: www4.ncsu.edu/unity/ lockers/users/f/felder/public/Papers/long5.html.
[19] OECD 2011a, Chart C2.1. Entry into postsecondary education is of course not the only obstacle to obtaining more STEM workers. Attention must also be paid to ensuring students complete their postsecondary education.
[20] College Board 2011, Table 7.
[21] Council of Economic Advisers 2011b, 1.
[22] National Center for Education Statistics 2010b, Figure 13 and Figure 3, 26 and 9.
[23] National Center for Education Statistics 2011, Figure 37, 46.
[24] Furchtgott-Roth, Joacobson, and Mokher 2009, 6.
[25] The Hamilton Project 2010, 3.
[26] Beede et al. 2011b, 4–5.
[27] Beede et al. 2011a, 3–5 and 7–8.
[28] Beede et al. 2011a, 5–6.
[29] See, for example, summary reports from Federal Reserve Board *Beige Book.*
[30] Partnership for a New American Economy 2011, 9.
[31] U.S. Department of Education 2011b.
[32] U.S. Department of Education 2011b.
[33] White House 2010.
[34] Council of Economic Advisers 2011a, 79.
[35] U.S. Department of Education, *Race to the Top Fund.*
[36] U.S. Department of Education 2011a.
[37] U.S. Department of Education 2011c.

End Notes for Infrastructure for the 21st Century

[1] Katz and Puentes 2008.
[2] Federal Communications Commission 2011a, 25.
[3] Pélissié du Rausas et al. 2011, 2.
[4] Pélissié du Rausas et al. 2011, 2.
[5] Pélissié du Rausas et al. 2011, 3.
[6] Pélissié du Rausas et al. 2011, 3.
[7] For a discussion of these concepts, see Appendix 2 in chapter 3.
[8] Edison Electric Institute 2011, 6.
[9] U.S. Department of Energy 2010.
[10] Council on Environmental Quality 2011.
[11] National Science and Technology Council 2011.
[12] Federal Communications Commission 2011a, 20.
[13] Economics and Statistics Administration and National Telecommunications and Information Administration 2011, 1.
[14] Household adoption is but one measure of how broadband contributes to a country's productivity. According to a study that takes account of a wider array of factors, such as use

of broadband by businesses and use of complementary technologies, the United States makes much more productive use of broadband than almost every other country in the world. See Council of Economic Advisers 2011, 67 and Waverman, et al. 2011.

[15] The National Broadband Map, a creation of the Department of Commerce's National Telecommunications and Information Administration (NTIA), depicts broadband availability for every community in the country; go to broadbandmap.gov.

[16] Federal Communications Commission 2011b.

[17] Among other projects, the Departments of Commerce and Agriculture invested about $7 billion in broadband projects under the American Recovery and Reinvestment Act. For more information, see BroadbandUSA at www2.ntia.doc.gov and Broadband Initiatives Program at www.rurdev.usda.gov/utp_bip.html.

[18] National Telecommunications and Information Administration 2011.

[19] For example, the Wireless Spectrum R&D (WSRD) Senior Steering Group coordinates spectrum-related R&D activities across the Federal government; for more information, see www.nitrd.gov/subcommittee/wireless spectrumrd.aspx

[20] Brown et al. 2011.

[21] Manyika and Roxburgh 2011, 6.

[22] Manyika et al. 2011.

[23] Manyika et al. 2011, 4–6 and 16.

[24] Brynjolfsson et al. 2011, 1.

[25] National Economic Council, *A Strategy for American Innovation,* 24.

[26] National Economic Council, *A Strategy for American Innovation,* 28.

[27] White House, *National Space Policy of the United States of America 2010,* 12.

[28] Federal Aviation Administration 2011.

[29] Federal Aviation Administration 2011, 5.

[30] Reilly 2011.

End Notes for Revitalizing Manufacturing

[1] Bureau of Economic Analysis, SURVEY OF CURRENT BUSINESS January 2011, Table 2.

[2] Bureau of Labor Statistics, 1961 to 2011, Table B–1.

[3] Executive Office of the President of the United States 2009, 7.

[4] Bureau of Economic Analysis, Trade in Goods and Services.

[5] PCAST *Report on Advanced Manufacturing* 2011.

[6] National Science Foundation, 2011. Detailed Statistical Tables NSF 11–301.

[7] Wolfe, Raymond M. 2009. Table 3. As cited by Tassey, 290.

[8] Tassey, 2010, 283–333.

[9] PCAST *Report on Advanced Manufacturing* 2011, 11.

[10] Bureau of Economic Analysis, GDP-by-Industry Data 2011.

[11] Board of Governors of the Federal Reserve System, Industrial Production and Capacity Utilization-G17 2011.

[12] Bureau of Labor Statistics, Labor Productivity and Costs, as cited by Executive Office of the President 2009, 6.

[13] Houseman, et. al. 2011, 115.

[14] Autor, Dorn and Hansen, 2011, 21.

[15] Bernard, Jensen and Schott 2004, 9.

[16] PCAST *Report on Advanced Manufacturing* 2011, 3.
[17] Council of Economic Advisers 2011, 95.
[18] Executive Office of the President of the United States 2009, 11.
[19] PCAST *Report on Advanced Manufacturing* 2011, 24.
[20] National Institutes of Standards and Technology, 2011, 1.
[21] PCAST *Report on Advanced Manufacturing* 2011, 17.
[22] National Center for Education Statistics 2010, Tables 196 and 198.
[23] National Center for Education Statistics 2010, Table 198.
[24] Lacey and Wright 2009, 88, Table 3.
[25] National Center for Education Statistics, IPEDS, 2010, Table 362.
[26] Executive Office of the President of the United States 2009, 3.
[27] PCAST *Report on Advanced Manufacturing* 2011, 20.

End Notes for The Private Sector as the Engine of Innovation

[1] The Obama Administration, through its *Strategy for American Innovation*, is committed to supporting these areas, as well as others. See *A Strategy for American Innovation: Securing Our Economic Growth and Prosperity,* www.whitehouse.gov/innovation/strategy.
[2] Porter 1998.
[3] Muro and Katz 2010.
[4] Estrin 2009.
[5] Porter 1998, 81–83.
[6] Delgado, Porter, and Stern 2011.
[7] Wheaton and Lewis 2002.
[8] Gibbs and Bernat 2001.
[9] Center for Economic Development and Business Research 2008.
[10] Greenstone and Moretti 2004; Greenstone, Hornbeck and Moretti, 2010.
[11] Haltiwanger, Jarmin, and Javier 2010.
[12] Ezell and Atkinson 2011, 26.
[13] National Economic Council 2011, *A Strategy for American Innovation*, 5 and 24.
[14] United States International Trade Commission 2007.
[15] Export Promotion Cabinet 2010, 4.
[16] Export Promotion Cabinet 2010, 4.
[17] The President's Economic Recovery Advisory Board 2010, 65.
[18] The President's Economic Recovery Advisory Board 2010, 65.
[19] Atkinson 2011.
[20] National Economic Council 2011, *A Strategy for American Innovation*, 42.
[21] National Economic Council 2011, *A Strategy for American Innovation*, 11.
[22] White House 2011a.
[23] National Academy of Sciences. National Academy of Engineering, and Institute of Medicine 2010, 57.
[24] White House 2011b.
[25] White House 2010.

In: Restoring American Competitiveness ... ISBN: 978-1-62081-509-0
Editors: A.J.Thomas and N.R.Ramirez

Chapter 2

AMERICA COMPETES 2010: FY2012 FUNDING AND FY2008-FY2011 FUNDING SUMMARY*

Heather B. Gonzalez

SUMMARY

The 112th Congress faces several budget and appropriations decisions that may affect implementation of the America COMPETES Reauthorization Act of 2010 (America COMPETES 2010, P.L. 111-358). Signed on January 4, 2011, this law seeks to improve U.S. competitiveness and innovation by authorizing, among other things, increased federal support for research and development (R&D) in the physical sciences and engineering, and science, technology, engineering, and mathematics (STEM) education. P.L. 111-358 reauthorizes the 2007 America COMPETES Act (America COMPETES 2007, P.L. 110-69), which enabled similar federal activities and programs from FY2008 to FY2010.

Overall, FY2012 enacted appropriations for programs with a specific funding level authorized by America COMPETES 2010 are close to FY2011 enacted levels. However, Congress shifted funding within the affected America COMPETES 2010 agencies in FY2012—particularly at the National Science Foundation (NSF) and National Institute of

* This is an edited, reformatted and augmented version of Congressional Research Service, Publication No. R41906, dated January 20, 2012.

Standards and Technology (NIST). These shifts increased funding for certain research accounts and decreased funding for other activities. Increases in research and construction funding at NSF were offset with reductions to STEM education; while at NIST, Congress did not appropriate funds for the Technology Innovation Program (TIP), among other things, and reduced the construction account. One deviation from the general trend of maintaining funding for COMPETES programs at or near FY2011 levels was at the Advanced Research Projects Agency-Energy (ARPA-E). Congress increased the ARPA-E budget to $275.0 million in FY2012, which is 53.1% more than the FY2011 level.

Some of the most closely watched R&D provisions in America COMPETES 2007 and 2010 are designed to double aggregate federal support for target accounts at the Department of Energy's (DOE) Office of Science, NIST, and NSF. The status of the so-called "doubling path" policy for R&D in the physical sciences and engineering is now uncertain. Prior-year appropriations for targeted doubling path accounts were generally below authorized levels and FY2012 enacted appropriations further reduce the average annual growth rate. At the FY2012 enacted rate, it would take almost 18 years to achieve the doubling. The Obama Administration's September 1, 2011, *Mid-Session Review* acknowledged that the doubling would need to be delayed.

The largest source of STEM education funding in America COMPETES 2010 is the Education and Human Resources (EHR) account at NSF. Congress reduced EHR funding from the FY2011 level in FY2012. Appropriators in both chambers also (1) rejected NSF's request to reduce funding for the Robert Noyce Teacher Scholarship and Mathematics and Science Partnership programs, (2) encouraged the Foundation to address the needs of Hispanic-Serving Institutions, and (3) directed NSF to distribute its report on best practices in STEM education. House appropriators raised concerns about certain DOE STEM education programs, suggesting that the department's fellowship programs may be more appropriate functions for NSF.

Other America COMPETES 2010-related provisions in FY2012 appropriations bills addressed the participation of under-represented minorities in STEM and supported regional economic development and manufacturing programs. **Table 1** at the end of this report summarizes the FY2012 appropriations status of selected provisions from the 2010 law.

INTRODUCTION

On January 4, 2011, President Barack Obama signed P.L. 111-358, the America COMPETES Reauthorization Act of 2010. The new law responds to concerns about U.S. competiveness by, among other things, increasing funding for research and development (R&D) in the physical sciences and engineering; and by authorizing certain federal science, technology, engineering, and mathematics (STEM) education programs. America COMPETES 2010 reauthorized selected provisions of the 2007 America COMPETES Act (P.L. 110-69).[1]

The purpose of this report is to provide information on the President's FY2012 budget request — and the status of FY2012 Congressional appropriations actions—for the agencies, programs, and activities authorized by America COMPETES 2010.[2] It also provides information on funding for the new law in FY2011 appropriations measures, and summarizes funding levels for the previous authorization period (FY2008-FY2010).

THE AMERICA COMPETES REAUTHORIZATION ACT OF 2010

Like the 2007 America COMPETES Act, America COMPETES 2010 is designed to "invest in innovation through research and development, to improve the competitiveness of the United States, and for other purposes."[3] In total, America COMPETES 2010 authorizes approximately $45.6 billion in funding between FY2010 and FY2013 for federal R&D in the physical sciences and engineering, STEM education, and other programs.

Among other things, the 2010 law increases funding authorizations for the National Science Foundation (NSF), National Institute of Standards and Technology (NIST) laboratories,[4] and the Department of Energy (DOE) Office of Science, and authorizes new technology transfer and commercialization activities at these agencies. It also authorizes inducement prizes at federal agencies; establishes a loan guarantee program for manufacturers; and establishes a Regional Innovation Program.

In STEM education, the 2010 reauthorization seeks to provide greater coordination of federal STEM education programs, authorizes support for academic programs that provide teacher certification concurrent with a bachelors degree in a STEM field, and repeals certain unfunded STEM education programs, among other things.

America COMPETES 2010 is an authorization measure. New programs—and funding increases for existing programs—authorized by the law will not be established or realized unless funded by an appropriations act.

FY2011 Funding for America COMPETES 2010

America COMPETES 2010 was enacted after the FY2011 appropriations and budget processes had begun and during an extended Congressional debate about the federal budget, deficit, and debt. Many of its FY2011 funding authorizations were not fully realized in FY2011 appropriations acts. This section provides an overview of FY2011 funding for the new law, including the status of so-called "doubling path" accounts and of selected other new and existing programs.

The Doubling Path

America COMPETES 2010 pursues a so-called "doubling path" policy for the NSF, NIST laboratories and construction, and the DOE Office of Science accounts (referred to herein as the "target accounts")—albeit at a reduced rate of average annual growth and over a longer period of time than was implicitly established in the original America COMPETES Act.[5] As authorized by the new law, the average annual growth rate in combined funding for target accounts is 6.3%. At this pace, the doubling period would be approximately 11 years (from a FY2006 baseline).

FY2011 enacted appropriations for target accounts were below the levels authorized by America COMPETES 2010. At the funding levels established by the Department of Defense and Full-Year Continuing Appropriations Act, 2011 (P.L. 112-10), the average annual growth rate in combined funding for target accounts was 4.6%. At this pace, the doubling period would be approximately 15 years (from a FY2006 baseline).[6]

New and Existing Programs

In general, new programs authorized by America COMPETES 2010 were not explicitly funded by FY2011 appropriations acts. One existing program, Teachers for a Competitive Tomorrow, was reduced to zero.[7] Other existing programs were generally held at FY2010 levels, minus a 0.2% across-the-board rescission.

Two existing programs, the Advanced Research Projects Agency-Energy (ARPA-E) at the DOE and the Manufacturing Extension Program (MEP) at NIST, were funded at higher levels in FY2011 than in FY2010. For ARPA-E, funding from regular FY2010 appropriations was zero.[8] Funding in FY2011 for ARPA-E was $179.6 million from regular appropriations, which represented the first regular appropriation for this program. For MEP, FY2011 appropriations were about 3.0% more than FY2010 levels ($128.4 and $124.7 million, respectively).

FY2008-FY2010 Funding for America COMPETES 2007

America COMPETES 2007 authorized $33.6 billion for federal programs and activities in R&D in the physical sciences and engineering and in STEM education, among other things, between FY2008 and FY2010. The law established new programs, such as the Math Now program; and reauthorized existing programs, such as the MEP program. The law also authorized increased funding levels for certain programs and activities, the largest of which were for the NSF, NIST laboratories, and the DOE's Office of Science. This section provides an overview of funding for the 2007 law, including the status of appropriations for the so-called "doubling path" policy and a summary of appropriations during the authorization period.

The Doubling Path

Actual appropriations for target accounts at the NSF, NIST, and DOE were less than the funding levels authorized by the 2007 law.[9] America COMPETES 2007 authorized a 10.3% average annual growth rate in funding for target accounts during the FY2008-FY2010 authorization period (from a FY2006 baseline). However, the actual average annual growth rate in appropriations during the authorization period was 6.4%. At that pace it would take about 11 years to achieve the doubling.[10]

Appropriations Summary

As with many authorization acts, America COMPETES 2007 was not fully funded and funding levels for its various provisions differed by agency, program, and fiscal year. In general, actual funding for the 2007 law was closest to authorized levels in 2009, when American Recovery and Reinvestment Act (ARRA, P.L. 111-5) funds augmented regular appropriations.[11] On an annual basis, appropriations during the law's authorization period may be summarized as follows:

- In **FY2008**, Congress provided funds for the Teachers for a Competitive Tomorrow program (at NSF) and Technology Innovation Program (at NIST).[12] Most of the law's other FY2008 authorizations went unfunded or were funded below authorized levels.[13]
- In **FY2009**, some America COMPETES Act programs were funded at authorized levels and others were not. For example, the NSF, NIST laboratories, and DOE's Office of Science were funded at or above authorized levels in FY2009 (including both regular and ARRA appropriations).[14] ARRA funds augmented regular appropriations for several America COMPETES Act programs in FY2009, including a $400 million additional investment in ARPA-E.[15] Other programs, including certain STEM education programs, received no funding.
- In **FY2010**, actual appropriations for P.L. 110-69 were generally below authorized levels. Actual appropriations were closest to authorized levels at NIST. Other than the Teachers for a Competitive Tomorrow program, STEM education programs at the Department of Education (ED) were not specifically funded.[16]

THE PRESIDENT'S FY2012 BUDGET REQUEST

The Administration's FY2012 budget request included funding for many existing programs and agencies reauthorized by America COMPETES 2010. For example, the Administration's FY2012 budget requests for the doubling path target accounts were near or above the levels authorized by America COMPETES 2010. However, many of the 2010 law's STEM education provisions were not explicitly funded in the Administration's FY2012 budget request or were funded below authorized levels. The FY2012 budget request also eliminated or merged certain STEM education programs at the Departments of Education and Energy, including programs authorized by America COMPETES 2010.

Of the new programs with specific funding authorizations in America COMPETES 2010, only the Regional Innovation Program (RIP) at the Department of Commerce was explicitly included in the Administration's

FY2012 budget request. The President's request for the RIP included funding for both the program as a whole, as well as a separate request for the Science Park Infrastructure Loan Guarantee component.

The following section discusses the President's FY2012 budget request for programs and agencies authorized by America COMPETES 2010.

Research

This section highlights the Administration's FY2012 requests for selected agencies and programs included in America COMPETES 2010 and examines the budgetary status of the doubling path target accounts.

Selected Agencies and Programs

President Obama's FY2012 budget request for the **NSF**'s Research and Related Activities (R&RA) account—which is the primary source of research funding at the Foundation—was $6.254 billion, or $20.0 million more than the authorized level of $6.234 billion. Within R&RA the President requested $160.5 million for the Experimental Program to Stimulate Competitive Research (EPSCoR) program and $9.2 million for the Partnerships for Innovation program. America COMPETES 2010 reauthorizes but does not specify funding levels for these programs.

The Administration's FY2012 budget request for the **DOE Office of Science** was $5.416 billion. This funding level was about $200.0 million less than the authorized level in America COMPETES 2010 ($5.614 billion). The Administration also sought $650.0 million for the ARPAE account at DOE,[17] $344.0 million more than the amount authorized in America COMPETES 2010 ($306.0 million).

At **NIST**, the Administration requested a total of $1.001 billion in FY2012, $30.2 million more than the authorized level of $970.8 million. Within the NIST budget the President requested $678.9 million, $17.8 million more than the authorized level of $661.1 million, for the Scientific and Technology Research and Services (STRS) account; and $84.6 million, $0.3 million less than authorized, for Construction of Research Facilities (CRF) account. Within the NIST Industrial Technology Services (ITS) account, the Administration requested $142.6 million for MEP, or $12.5 million less than the $155.1 million authorized; and $7.7 million for the Malcolm Baldrige Quality Award program, $2.6 million less than the authorized amount of $10.3

million. The Administration did not specifically request funds for the NIST Green Jobs Act (Title VII of America COMPETES 2010).

The Doubling Path

At the time of its release in February 2011, the FY2012 *President's Plan for Science and Innovation* stated that the FY2012 budget request "maintains the President's commitment to double the budgets of three key science agencies"[18] (i.e., the target accounts at NSF, NIST laboratories and construction, and the DOE Office of Science). However, the Administration's September 1 *Mid-Session Review* acknowledged that the doubling goal would need to be delayed.

STEM Education

The President's FY2012 budget proposal for STEM education provisions in America COMPETES 2010 included funding requests for existing programs at NSF, and funding and program changes at the Departments of Education and Energy.[19] The Administration's FY2012 budget did not include specific requests for new STEM education programs authorized by America COMPETES 2010, such as the STEM-Training Grant Program.

The primary source of funding for STEM education activities at NSF is the Education and Human Resources (E&HR) account.[20] The Administration requested $911.2 million for E&HR in FY2012. This amount was $67.8 million (6.9%) less than the America COMPETES 2010 authorized level of $979.0 million and $50.2 million (5.8%) more than the FY2011 enacted level of $861.0 million. The NSF budget request highlighted proposed FY2012 increases for the Graduate Research Fellowship program ($198.1 million), which would have funded 2,000 new awards; the Advanced Technological Education program ($64.0 million), which focuses on technical education at community colleges; and the new Widening Implementation and Demonstration of Evidence-based Reforms (WIDER) program ($20.0 million), which sought to broadly implement undergraduate STEM education reforms.

The FY2012 NSF budget request also included funding for existing STEM education programs authorized under America COMPETES 2010, but for which the act does not specify funding levels. These include the Integrative Graduate Education and Research Traineeship (IGERT), the Robert Noyce Teacher Scholarship (Noyce) program, Research Experiences for

Undergraduates (REU), and the STEM Talent Expansion Program (STEP), among others. The Administration's FY2012 requests for these programs were $62.5 million for IGERT, $45.0 million for Noyce, $66.0 million for REU, and $35.5 million for STEP.

Both America COMPETES 2007 and 2010 include provisions authorizing a program to support Hispanic-serving institutions (HSIs) at the NSF. Section 7033 of America COMPETES 2007 directed the Foundation to establish a program for HSIs. Section 512 of America COMPETES 2010 directs the NSF to maintain its HSI program—and all other similar programs, such as the Historically Black Colleges and Universities Undergraduate Program (HBCU-UP)—as separate programs.[21] As directed by the 2010 law, the Administration's FY2012 budget request for NSF maintained the Foundation's minority-serving institution programs as separate programs, albeit within a new "framework" called Broadening Participation at the Core (BPAC). However, the Foundation has not established an HSI-specific program—either within or apart from the BPAC framework—as directed by either America COMPETES 2007 or 2010. NSF's FY2012 budget incorporated a $20.0 million request for a new program—Transforming Broadening Participation through STEM (TBPS)—that would have included "increased engagement with Hispanic-serving institutions"[22] in its mission.

The President's FY2012 budget request for ED reorganized the department (as previously proposed in the FY2011 request).[23] The proposed reorganization eliminated and consolidated certain programs, including America COMPETES 2010 programs.[24] For example, under the reorganization plan, both the Teachers for a Competitive Tomorrow (TCT) and Advanced Placement (AP) programs would have been eliminated and their program functions absorbed into the newly created Teacher and Leader Pathways and College Pathways and Accelerated Learning programs, respectively.[25] The FY2012 request for Teacher and Leader Pathways was $250.0 million.[26] The FY2012 request for College and Accelerated Learning Pathways was $86.0 million.[27]

Similarly, the Administration's FY2012 request for DOE proposed both funding for, and changes to, the department's STEM education programs. DOE typically does not request specific funding for America COMPETES Act-authorized STEM education programs. However, the department says it operates programs that correspond with its responsibilities under the law.[28] According to DOE, the Academies Creating Teacher Scientists (DOE ACTS) program corresponded with the Summer Institutes program from America COMPETES 2007 and 2010.[29] In accordance with the recommendations of a 2010 DOE Committee of Visitors report, the Administration sought to terminate DOE ACTS in FY2012.[30] DOE also asserts that the Computational Science Graduate Fellowship and the Office of Science Graduate Fellowship programs are consistent with the Protecting America's Competitive Edge (PACE) graduate fellowship program authorized by both America COMPETES 2007 and the 2010 reauthorization.[31] The Administration's FY2012 requests for the Computational Science and Office of Science graduate fellowship programs were $6.0 and $16.1 million, respectively.

Other Provisions

The President's FY2012 budget request contained funding for other America COMPETES 2010 provisions as well—including $40.0 million for the new RIP program and $7.0 million for a new Science Park Infrastructure Loan Guarantee program at the Department of Commerce's Economic Development Administration (EDA).[32] The Administration's FY2012 budget request did not include specific funding for the new Federal Loan Guarantees for Innovative Technologies in Manufacturing program at the Department of Commerce[33] or for the NIST Green Jobs Act of 2010, both of which were authorized by America COMPETES 2010.[34]

FY2012 Congressional Action

Funding for America COMPETES 2010 programs and agencies is typically included in three appropriations acts: [35]

- *Commerce, Justice, Science, and Related Agencies* (CJS), for NSF, NIST, and other Department of Commerce programs;[36]
- *Energy and Water Development* (Energy-Water), for DOE programs;[37]
- *Labor, Health and Human Services, Education, and Related Agencies* (LaborHHS-Education), for ED programs.

As appropriations measures typically include a variety of provisions and programs, this section focuses on funding provisions that relate most closely to policies, programs, agencies, and activities specifically authorized by America COMPETES 2010. **Table 1** summarizes the FY2012 appropriations status of these selected provisions.

Commerce, Justice, Science, and Related Agencies (CJS)

On November 18, 2011, the President signed H.R. 2112 (Consolidated and Further Continuing Appropriations Act, 2012). The bill, which became P.L. 112-55, provides FY2012 appropriations for the DOC, NIST, NSF, and other CJS agencies.

This section compares funding levels in P.L. 112-55—and the conference report that accompanied H.R. 2112 (H.Rept. 112-284)—with selected FY2011 enacted appropriations, FY2012 Administration budget requests, and FY2012 America COMPETES 2010 funding authorizations. Policy provisions in H.Rept. 112-284 that relate to specific America COMPETES 2010 authorizations are also noted herein. Additionally, this section includes selected directives and language from House and Senate appropriations committee reports on CJS FY2012 funding bills. These provisions may be found in H.Rept. 112-169 and S.Rept. 112-78.

Department of Commerce

NIST

P.L. 112-55 provides $750.8 million in funding for NIST. This amount is $0.7 million (0.1%) more than the FY2011 enacted level, $250.3 million (25.0%) less than the Administration's FY2012 request, and $227.0 million (23.2%) less than the America COMPETES authorized level. Although total FY2012 funding for NIST is similar to the FY2011 enacted level, P.L. 112-55 increases the main research account by 11.8%. This increase is offset by, among other things, a 20.7% reduction ($14.5 million) in the main construction account (relative to FY2011 enacted funding) and termination of the TIP program, which was first authorized by America COMPETES 2007. P.L. 112-55 funds the MEP program, which was reauthorized by America COMPETES 2010, at $128.4 million in FY2012. This amount is consistent with FY2011 enacted appropriations for the MEP program.

Regional Innovation Program and Innovative Technologies in Manufacturing

America COMPETES 2010 authorized two regional economic development programs at the EDA: RIP, which includes funding for loan guarantees for science parks, and the Federal Loan Guarantees for Innovative Technologies in Manufacturing program (hereinafter called the "manufacturing loan guarantee program"). P.L. 112-55 provides up to $5.0

million each for the RIP and manufacturing loan guarantee programs and caps the total guaranteed loan principal for both programs (combined) at $70.0 million.

FY2012 enacted funding for the RIP program is $35.0 million less than the Administration's request for $40.0 million. P.L. 112-55 also specifically authorizes funding for provisions of the RIP program that would establish a science park loan guarantee program at EDA. These provisions had previously been rejected by the Senate Committee on Appropriations (S.Rept. 112-78.), which stated that it did not support the Administration's request to make the EDA a loan-generating agency. The conference report on H.R. 2112 encourages EDA to support regional innovation activities from within the Economic Adjustment Assistance (EAA) program.

The Administration did not request funding for the manufacturing loan guarantee program in FY2012, but the funding level in P.L. 112-55 is consistent with the House Committee on Appropriations' recommendation in H.Rept. 112-169. H.Rept. 112-169 also encourages the EDA to give consideration to applicants from areas with high unemployment, directs EDA to use EAA funds to support the program, requires the EDA to submit a report on the program with its FY2013 budget request, and directs EDA to consult with the Inspector General on oversight and other issues before implementing the program.

National Science Foundation

Top Line Allocations and Changes

P.L. 112-55 increases funding for NSF by 2.5%—relative to FY2011 levels—to a total of $7,033.1 million in FY2012. This amount is $733.9 million (9.4%) less than the Administration's request and $766.9 million (9.8%) less than the America COMPETES 2010 authorized funding level. P.L. 112-55 increases NSF's main research and construction accounts and reduces the Foundation's main education account (compared to FY2011 levels). Additionally, P.L. 112-55 allows the NSF to shift up to $50.0 million from the research account to the construction account and provides a small (1.6%) increase to the NSF Inspector General and a small reduction (2.0%) to the National Science Board budget.

Research Provisions

The primary source of research funding at the NSF is the Research and Related Activities (RRA) account. P.L. 112-55 increases funding for the RRA account by $155.1 million (2.8%) over the FY2011 enacted level to a total of $5,719.0 in FY2012. This amount is $534.5 million (8.5%) less than the Administration's budget request and $515.3 million (8.3%) less than the America COMPETES authorized level. P.L. 112-55 also sets aside $150.9 million of the RRA allocation for the EPSCoR program, allows the NSF to transfer up to $50.0 million from RRA to the Foundation's main construction account, and permits the NSF to use RRA funds to reimburse other federal agencies for support of the U.S. Antarctic program.

The conference report on H.R. 2112 endorses most Administration-proposed reductions to RRA programs in FY2012. It also adopts language from H.Rept. 112-169 that supports planned NSF activities in advanced manufacturing and agrees to language from S.Rept. 112-78 that provides $165.6 million for cybersecurity research. Other America COMPETES-related provisions in FY2012 CJS appropriations reports commend NSF for launching the new Innovation Corps (I-CORPS) program, which is designed to bridge the gap between laboratory research activities and technology commercialization (S.Rept. 112-78); and direct the Foundation to report on its plans for offering innovation prizes (H.Rept. 112-169). Provisions in S.Rept. 112-78 support multidisciplinary, high-risk research and investments in robotics; while H.Rept. 112-169 directs the NSF to prioritize materials research and disciplinary and interdisciplinary research in the natural and physical sciences, math, and engineering. H.Rept. 112-169 also requires the NSF to report on steps that can be taken to balance the imperatives of public access to, and protection of, data.

STEM Education

Although support for STEM education activities may be found across the NSF, the primary source of funding for these programs is the Education and Human Resources (EHR) account. FY2012 funding for EHR is $829.0 million. This amount is $32.0 million (3.7%) less than the FY2011 enacted level, $82.2 (9.0%) less than the Administration's request, and $150.0 million (15.3%) less than the America COMPETES 2010 authorized level. P.L. 112-55 also sets aside $54.9 million of EHR's total FY2012 allocation for the Noyce program. This amount is consistent with FY2011 enacted funding for the Noyce program.

The conference report on H.R. 2112 endorses the Administration's proposed terminations and reductions in EHR—except for proposed reductions to the Mathematics and Science Partnership and Noyce programs. Both H.Rept. 112-169 and S.Rept. 112-78 urge the NSF to ensure that GRF applicants are not rejected for reasons unrelated to the merits of their proposed research (e.g., the applicant's major). Among the provisions in S.Rept. 112-78 are those expressing support for the Informal Science Education program and those that strongly encourage the Foundation to continue supporting both the Professional Science Master's program and undergraduate science and engineering education. H.Rept. 112-169, on the other hand, includes language supporting Administration-proposed cuts to certain STEM education programs and encouraging the Foundation to continue cooperating with federal auditors to identify duplication in the federal STEM education portfolio.

Broadening Participation

The conference report on H.R. 2112 provides FY2011 funding levels for NSF's Broadening Participation at the Core programs (e.g., the Tribal Colleges and Universities Program, Historically Black Colleges and Universities Undergraduate Program) and directs the NSF to report on how it will address the needs of Hispanic-Serving Institutions in FY2012.

Energy and Water Development (Energy-Water)

The President signed H.R. 2055 (The Military Construction and Veterans Affairs and Related Agencies Appropriations Act, 2012) on December 23, 2011. The bill, which became P.L. 112-74, contains the content of nine regular appropriations acts, including the Energy and Water Development Appropriations Act, 2012. H.Rept. 112-331 accompanied H.R. 2055 when it was reported from conference and contains additional provisions and explanations. The House approved H.R. 2354 (Energy and Water Development Appropriations, 2012) on July 15, 2011. H.Rept. 112-118 accompanied H.R. 2354 when it left the House Committee on Appropriations and included additional explanations and provisions. The Senate Committee on Appropriations amended and reported on H.R. 2354 on September 7, 2011. S.Rept. 112-75 accompanied H.R. 2354 when it left the Senate Committee on Appropriations and contained additional explanations and provisions.

This section compares selected funding levels in P.L. 112-74 with FY2011 enacted appropriations, FY2012 Administration budget requests, and FY2012 America COMPETES 2010 funding authorizations. It also notes policy provisions in the House and Senate appropriations committee reports (H.Rept.

112-118 and S.Rept. 112-75, respectively) and the conference report (H.Rept. 112-331) that relate to specific America COMPETES 2010 authorizations.

Department of Energy

Office of Science (SC)

P.L. 112-74 provides $4.889 billion for the Office of Science in FY2012. This amount is $46.3 million (1.0%) more than the FY2011 enacted level of $4.843 billion. It is $527.1 million (9.7%) less than the Administration's FY2012 request for $5.416 billion and $725.0 million (12.9%) less than the America COMPETES 2010 authorized funding level of $5.614 billion.

Program Evaluation

Among other things, the conference report directs DOE to create a performance ranking of all ongoing multi-year research projects across the Office of Science's six major research programs. Specifically, DOE must compare current performance with original project goals and prepare a report that includes a list of the total number of, and amount of funding for, multi-year awards that were terminated before their terms ended in both FY2011 and FY2012. SC's program evaluation practices are also a central concern in the House and Senate committee reports. Both reports state that most Office of Science research should have concrete goals and measurable performance. However, the Senate report directs SC to evaluate and rank all ongoing multi-year research projects across its portfolio while the House report directs the office to establish project rankings and terminate low-performing projects in the Office of Basic Energy Sciences (BES). (The conference report is consistent with the Senate report.) The Senate report also 1) commends SC for identifying clear priorities for the *outcomes* of basic research and encourages SC to establish an advisory committee to help prioritize among different *topical areas* of research; and 2) directs BES to implement the recommendations of the April 2010 Basic Energy Sciences Advisory Committee report on ways to strengthen the link between basic research and industry.

Multi-Year Awards

The conference report directs DOE to transition to a funding model in which it fully funds multi-year awards with appropriated funds each year (with some exceptions for items such as large construction projects).

These provisions are consistent with similar directives from the House committee report which, among other things, specifically critiques multi-year funding practices at the Office of Science.

In particular, the House committee report rejects the way SC funds multi-year projects (partially each year, rather than so-called "up-front full funding") and notes that while this funding model may allow the department to support more projects in a given fiscal year, it assumes funds that do not exist (and may not materialize) and limits the department's ability to adjust its programs from year-to-year.

STEM Education

The conference report provides $5.0 million for SC's graduate fellowship program in FY2012. These funds are for the cohort established in FY2010. The House report includes language prohibiting DOE from funding fellowship and scholarship programs in FY2012 unless those programs are included in the Department's FY2012 budget request or are supported in H.R. 2354 and directs DOE to provide the committee with a comprehensive list of all educational activities funded in FY2012.

Other STEM education-related provisions in the House report include a directive requiring DOE to prepare a 10-year plan for the fellowship program including objectives, proposed funding, and an explanation justifying SC support for the fellowship program "when other agencies, in particular the National Science Foundation, are the primary federal entities for such purposes."[38]

The Senate report, among other things, recommends $7.5 million for SC's graduate fellowship program. It also encourages the Office of Science to redirect funds from STEM education program terminations to the unfunded Distinguished Scientist program as authorized by America COMPETES 2007 and 2010.

Advanced Research Projects Agency—Energy (ARPA-E)

P.L. 112-74 provides $275.0 million for ARPA-E in FY2012. This amount is $95.4 million (53.1%) more than the FY2011 enacted level of $179.6 million, $375.0 million (57.7%) less than the Administration's FY2012 request for $650.0 million,[39] and $31.0 million (10.1%) less than the America COMPETES 2010 authorized funding level of $306.0 million.

The conference report directs DOE to report on how it has or will implement in all programs certain ARPA-E management practices perceived as successful and highlighted by the President's Council of Advisors on Science and Technology (PCAST).

These include the ARPA-E review process, the speed with which ARPA-E negotiates contracts and grants, co-location of support functions (e.g., procurement, information technology) in program offices, and workforce.

Broadening Participation

The conference report adopts language from the House report that encourages DOE to engage in competitive grant programs that increase the number of underrepresented college minorities in STEM fields and to report on the effectiveness of these efforts.

The Senate report encourages DOE to continue supporting the research and development needs of Historically Black Colleges and Universities.

Labor, Health and Human Services, Education, and Related Agencies (Labor-HHS-Education)

The President signed H.R. 2055 (The Military Construction and Veterans Affairs and Related Agencies Appropriations Act, 2012) on December 23, 2011. The bill, which became P.L. 112-74, contains the content of nine regular appropriations acts, including the Departments of Labor, Health and Human Services, and Education, and Related Agencies Appropriations Act, 2012. H.Rept. 112-331 accompanied H.R. 2055 when it was reported from conference and contains additional provisions and explanations.

The Senate Committee on Appropriations approved S. 1599 (Departments of Labor, Health and Human Services, and Education, and Related Agencies Appropriations Act, 2012) on September 21, 2011. S.Rept. 112-84 accompanied S. 1599 when it left the Senate Committee on Appropriations and included additional explanations and provisions.

Neither the House Committee on Appropriations, nor the full House or Senate, approved a stand-alone Labor-HHSAppropriations measure.

This section compares selected funding levels in P.L. 112-74—and the conference report that accompanied H.R. 2055 (H.Rept. 112-331)—with FY2011 enacted appropriations, FY2012 Administration budget requests, and FY2012 America COMPETES 2010 funding authorizations. Policy provisions in S.Rept. 112-84 that relate to specific America COMPETES 2010 authorizations are also noted herein.

Department of Education (ED)

Teachers for a Competitive Tomorrow (TCT)

Although America COMPETES 2010 reauthorizes the Teachers for a Competitive Tomorrow program and authorizes $4.0 million per year for TCT in FY2011, FY2012, and FY2013, P.L. 112-10 (Department of Defense and Full-Year Continuing Appropriations Act, 2011) eliminated funding for the program in FY2011. H.Rept. 112-331 does not specify funding for TCT programs in FY2012. The future of the TCT program is unclear.

Advanced Placement (AP)

P.L. 112-74 provides $27.0 million for AP programs at ED—$16.3 million (37.6%) less than the FY2011 enacted amount of $43.3 million and $31.7 million (42.3%) less than the America COMPETES 2010 authorized funding level of $75.0 million.

As part of its plan to re-organize the Department of Education, the Administration did not ask for separate funding for AP programs in FY2012.

Although Congress provided funding for AP programs in FY2012, the AP programs it funded were not authorized by America COMPETES 2010.

At least two federal laws authorize AP programs at ED—America COMPETES 2010 and the Elementary and Secondary Education Act of 1965, as amended by No Child Left Behind (ESEA).

Advanced Placement program provisions in ESEA differ from similar provisions in America COMPETES 2010—for example, the ESEA program prioritizes certain AP incentive program grant applications, while the America COMPETES 2010 program prioritizes others—though the practical impact of these distinctions on program implementation is unclear.

Both P.L. 112-74 and ED's FY2012 budget documents refer to ESEA for AP program and funding authority.

The Senate Committee on Appropriations' recommendation for AP programs in FY2012 is $43.3 million (S.Rept. 112-84).

Alignment of Education Programs

P.L. 112-74 does not provide specified funding for alignment of education programs.[40]

America COMPETES 2010 authorizes $120.0 million for these activities in each fiscal year from FY2011 to FY2013.

Table 1. America COMPETES Reauthorization Act of 2010 (P.L. 111-358): Selected Programs and FY2012 Appropriations Status (in millions of dollars)

Programs	FY2011 Appropriation (P.L. 112-10)	FY2012 Authorization (P.L. 111-358)	FY2012 Request	House	Senate	FY2012 Final (P.L. 112-74)
	Department of Education					
Teachers for a Competitive Tomorrow – Baccalaureate (Sec. 1003)	$0.0a	$2.0	$0.0b	n/a	n/d	$0.0
Teachers for a Competitive Tomorrow – Master's (Sec. 1003)	$0.0a	$2.0	$0.0b	n/a	n/d	$0.0
Advanced Placement and International Baccalaureate Programs (Sec. 1003)	n/d	$75.0	$0.0c	n/a	n/ac	n/ac
Alignment of Education Programs (Sec. 1003)	n/d	$120.0	n/dd	n/a	n/d	n/d
	Department of Energy					
Summer Institutes (Sec. 901)	n/d	$25.0	$0.0e	n/d	n/d	n/d
Nuclear Science Program Expansion Grants for Institutions of Higher Education (Sec. 902)	n/d	$10.1	n/d	n/d	n/d	n/d
Nuclear Science Competitiveness Grants for Institutions of Higher Education (Sec. 902)	n/d	$8.5	n/d	n/d	n/d	n/d

Table 1. (Continued)

Programs	FY2011 Appropriation (P.L. 112-10)	FY2012 Authorization (P.L. 111-358)	FY2012 Request	House	Senate	FY2012 Final (P.L. 112-74)
Hydrocarbon Systems Science Talent Program (Expansion Grants, Sec. 902)	n/d	$10.1	n/d	n/d	n/d	n/d
Early Career Awards (Sec. 902)	n/d	$25.0	n/df	n/d	n/d	n/d
Protecting America's Competitive Edge (PACE) Graduate Fellowship Program (Sec. 902)	n/d	$21.2	n/dg	n/d	n/dh	n/dh
Distinguished Scientist Program (Sec. 902)	n/d	$32.0	$0.0i	n/d	n/dj	n/dj
Basic Research (Office of Science, Sec. 903)	$4,842.7	$5,614.0	$5,416.1	$4,800.0	$4,842.7	$4,889.0
Advanced Research Projects Agency—Energy (Sec. 904)	$179.6	$306.0	$650.0k	$179.6	$250.0	$275.0
	Department of Commerce					
Federal Loan Guarantees for Innovative Technologies in Manufacturing (New, Sec. 602)	n/d	$20.0	n/d	$5.0	n/d	$5.0
Regional Innovation Program (New, Sec. 603)	$0.0	$100.0	$40.0	n/d	n/dl	
Loan Guarantees for Science Park Infrastructure (New, Sec. 603)	n/d	$7.0	$7.0	n/d	n/dl	$5.0m

Programs	FY2011 Appropriation (P.L. 112-10)	FY2012 Authorization (P.L. 111-358)	FY2012 Request	House	Senate	FY2012 Final (P.L. 112-74)
	National Institute of Standards and Technology					
Total	$750.1	$977.8	$1,001.1	$700.1	$680.0	$750.8
Scientific & Technical Research & Services	$507.0	$661.1	$678.9	$516.5	$500.0	$567.0
Construction of Research Facilities	$69.9	$84.9	$84.6	$55.3	$60.0	$55.4
Industrial Technology Services	$173.3	$224.8	$237.6	$128.3	$120.0	$128.4
Manufacturing Extension Partnership	$128.4	$155.1	$142.6	$128.3	$120.0	$128.4
Baldrige Performance Excellence Program	$9.6	$10.3	$7.7	$0.0	$0.0	$0.0
NIST Green Jobs Act of 2010 (New, Sec. 703)	n/d	$7.0	n/d	n/d	n/d	n/d
	National Science Foundation					
Total	$6859.9	$7,800.0	$7,767.0	$6,853.0	$6,698.1	$7,033.1
Research & Related Activities	$5,563.9	$6,234.3	$6,253.5	$5,601.4	$5,443.0	$5,719.0
Education & Human Resources	$861.0	$979.0	$911.2	$834.2	$829.0	$829.0
Major Research Equipment and Facilities Construction	$117.1	$225.5	$224.7	$99.9	$117.1n	$167.1
Agency Operations & Award Management	$299.4	$341.7	$357.7	$299.1	$290.4	$299.4
National Science Board	$4.5	$4.8	$4.8	$4.5	$4.4	$4.4
Office of the Inspector General	$14.0	$14.7	$15.0	$14.0	$14.2	$14.2
STEM-Training Grant Program (New, Sec. 556)	n/d	$10.0	n/d	n/d	n/d	n/d

Sources: U.S. Department of Commerce, Economic Development Administration, *FY2012 Congressional Budget Request*, no date, http://www.osec.doc.gov/bmi /budget/12CJ/EDA_FY_2012_Congressional_Submission.pdf; U.S. Department of Education, *FY 2012 Department of Education Justifications of Appropriation Estimates to the Congress*, March 31, 2011, http://www2.ed.gov/about /overview/budget/budget12/justifications/index.html; U.S. Department of Energy, *FY2012 Congressional Budget Request: Budget Highlights*, February 2012 [sic], http://www.cfo.doe.gov/ budget/12budget/Content/FY2012Highlights.pdf; National Institute of Standards and Technology, *Fiscal Year 2012 Budget Submission to Congress*, January 28, 2010, http://www.osec.doc.gov/bmi/budget /12CJ/2012_NIST_& _NTIS_Cong_Budget.pdf; National Science Foundation, *NSF FY 2012 Budget Request to Congress: Overview*, no date, http://www.nsf.gov /about/budget/fy2012/pdf/01_fy2012.pdf; Department of Defense and Full-Year Continuing Appropriations Act, 2011 (P.L. 112-10), Consolidated Appropriations Act, 2010 (P.L. 111-17), Energy and Water Development and Related Agencies Appropriations Act, 2010 (P.L. 111-85), America COMPETES Reauthorization Act of 2010 (P.L. 111-358), and e-mail and telephone communication between the author and Patricia Temple, U.S. Department of Energy, April 14, 2011. Funding levels in the "House" column are from H.R. 2354, as passed by the House, and H.Rept. 112-118; and from H.R. 2596, as reported by the House Committee on Appropriations, and H.Rept. 112-169. Funding levels in the "Senate" column are from H.R. 2112, as passed by the Senate, and S.Rept. 112-78; H.R. 2354, as reported by the Senate Committee on Appropriations, and S.Rept. 112-75; and S. 1599, as reported by the Senate Committee on Appropriations, and S.Rept. 112-84. Funding levels in the column titled "FY2012 Enacted" are from P.L. 112-55 and H.Rept. 112-284, and P.L. 112-74 and H.Rept. 112-331.

Notes: n/d = not defined; CRS was unable to identify a specific, defined appropriation or budget request for the authorization. n/a = not applicable. Funding levels for FY2011 include 0.2% across-the-board rescission as per P.L. 112-10, Sec. 119. Programs designated as "new" are new to the America COMPETES Act. Totals may not add due to rounding.

[a] P.L. 112-10 eliminated funding for the Teachers for a Competitive Tomorrow program in FY2011.

[b] The FY2012 budget request for ED merges these programs into the proposed new program Teacher and Leader Pathways (TLP). The FY2012 request for TLP was $250.0 million.

[c] The Administration's FY2012 request would merge Advanced Placement programs into the proposed new program College Pathways and Accelerated Learning (CPAL). The FY2012 request for CPAL was $86.0 million. FY2012 funding for AP programs (as authorized by ESEA, not COMPETES) in P.L. 112-74 is $27.0 million. S.Rept. 112-84 recommends $43.3 million for AP programs in FY2012.

[d] ED has similar programs and activities but does not rely on P.L. 111-358 for general statutory authority to undertake alignment activities. The exception to this rule is for state education data system elements, for which ED relies on P.L. 110-69, Sec. 6401.

[e] According to DOE, this program corresponds with the DOE ACTS program. DOE's FY2012 request proposed elimination of DOE ACTS.

[f] According to DOE, there is no line item for Early Career Awards but funds were requested for the program in both FY2011 and FY2012. The agency initiated the program in FY2012 with American Recovery and Reinvestment Act (ARRA) funds.

[g] According to DOE, the department manages at least two programs that are consistent with PACE provisions: (1) the Computational Science Graduate Fellowship (CSGF) in the Office of Science, Advanced Scientific Computing Research, and (2) the Graduate Fellowship (SCGF) program in the Office of Science, Workforce Development for Teachers and Scientists. The FY2012 request

for CSGF was $6.0 million and the FY2012 request for SCGF was $16.1 million. H.Rept. 112-331 provides $5.0 million for the SCGF program in FY2012.

[h] S.Rept. 112-75 recommends $7.5 million for the SCGF in FY2012.

[i] According to DOE, no funds have been requested or appropriated for this activity.

[j] S.Rept. 112-75 encourages DOE to redirect Office of Science, Workforce Development for Teachers and Scientists' funding from programs proposed for termination to the Distinguished Scientist program.

[k] The FY2012 request for ARPA-E included $100.0 million from the Wireless Innovation (WIN) Fund.

[l] Although H.R. 2112, as passed by the Senate, does not specify funding for the RIP program (or the loan guarantees for science parks components of that program), provisions of the bill state that funding for the DOC's Economic Development Administration includes support for the RIP as authorized by America COMPETES 2010. S.Rept. 112-78 also includes language stating that the Committee does not support the Administration's request to make EDA a loan-generating agency (S.Rept. 112-78, p. 13).

[m] P.L. 112-55 provides up to $5.0 million for *both* RIP program grants and science park loan guarantees. H.Rept. 112-284 also encourages EDA to support regional innovation activities (more generally) within the EAA account.

[n] H.R. 2112, as passed by the Senate, would permit the NSF to transfer up to $100.0 million from the RRA account to the Major Research Equipment and Facilities Construction account.

End Notes

[1] This report refers to the America COMPETES Reauthorization Act of 2010 as "America COMPETES 2010" and to the America COMPETES Act as "America COMPETES 2007." The full title of the America COMPETES Act is the "America Creating Opportunities to Meaningfully Promote Excellence in Technology, Education, and Science Act."

[2] Numbers reported are rounded, therefore small inconsistencies may occur in some cases.

[3] P.L. 111-358, Purpose.

[4] NIST is part of the U.S. Department of Commerce.

[5] As authorized in America COMPETES 2007, the average annual growth rate in combined funding for the target accounts was 10.3%. At this pace, the doubling period would be approximately seven years (from a FY2006 baseline). Actual appropriations to the target accounts over the FY2008-FY2010 authorization period grew at a pace of 6.4%. At this rate of average annual growth, the doubling period would be approximately 11 years (from a FY2006 baseline).

[6] For more information on the status of the doubling path effort, see CRS Report R41951, *An Analysis of Efforts to Double Federal Funding for Physical Sciences and Engineering Research*, by John F. Sargent Jr.

[7] P.L. 112-10, Title VIII, Sec. 1842.

[8] However, ARPA-E received about $415.0 million in FY2009, mostly from the American Recovery and Reinvestment Act (P.L. 111-5).

9 In 2006 President George W. Bush proposed a doubling path policy as part of the *American Competitiveness Initiative* (ACI). The President's proposal would have increased funding for the NSF, NIST core, and DOE Office of Science over a 10-year period (from a FY2006 baseline). See, U.S. President (George W. Bush), Executive Office of the President, Office of Science and Technology Policy, *American Competitiveness Initiative*, 2006, http://www.whitehouse.gov/files/documents/ostp/pdf/1pger_aci.pdf.

[10] For more information on the status of the doubling path effort, see CRS Report R41951, *An Analysis of Efforts to Double Federal Funding for Physical Sciences and Engineering Research*, by John F. Sargent Jr.

[11] In the America COMPETES Act context, most analysts have treated ARRA funds as a one-time supplement, not as an agency or program baseline.

[12] The Technology Innovation Program replaced the Advanced Technology Program.

13 This may be attributable, at least in part, to the fact that P.L. 110-69 passed after much of the FY2008 appropriations process had already taken place.

[14] Excluding ARRA funds, the NSF and DOE Office of Science were funded below FY2009 authorized levels. The target accounts at NIST—e.g., the Scientific and Technology Research and Services (STRS) and the Construction of Research Facilities (CRF) accounts—were funded above authorized levels in regular appropriations acts *and* received supplemental funding from ARRA.

[15] For more information, see CRS Report RL34396, *The America COMPETES Act and the FY2009 Budget*, by Deborah D. Stine.

[16] For more information, see CRS Report R40519, *America COMPETES Act and the FY2010 Budget*, by John F. Sargent Jr.

[17] This amount includes $550.0 million in regular appropriations and $100.0 million from the Administration's proposed Wireless Innovation (WIN) fund.

[18] U.S. President Barack Obama, Executive Office of the President, Office of Science and Technology Policy, "The President's Plan for Science and Innovation: Doubling Funding

for Key Science Agencies in the 2012 Budget," press release, February 14, 2011, http://www.whitehouse.gov/sites/default/files/microsites/ostp/FY12-doubling-fs.pdf.

[19] See Executive Office of the President, Office of Science and Technology Policy, "Winning the Race to Educate Our Children," press release, February 14, 2011, http://www.whitehouse.gov/sites/default/files/microsites/ostp/FY12- STEM-fs.pdf.

[20] The NSF Research and Related Activities account also supports some STEM education activities.

[21] The Foundation previously proposed consolidating its minority-serving institution programs. Congressional authorizers and appropriators both rejected that proposal.

[22] National Science Foundation, *NSF FY 2012 Budget Request to Congress*, February 14, 2011, p. EHR-19, http://www.nsf.gov/about/budget/fy2012/pdf/01_fy2012.pdf.

[23] Congress must authorize this reorganization for it to take effect. The FY2011 appropriation to ED retained the existing department structure and organization. Legislative debate about the President's proposal has continued in the context of the proposed reauthorization of the Elementary and Secondary Education Act of 1965, as amended by the No Child Left Behind Act. That debate began in the 111th Congress and continues in the 112th.

[24] For more information, see CRS Report R41355, *Administration's Proposal to Reauthorize the Elementary and Secondary Education Act: Comparison to Current Law*, by Rebecca R. Skinner et al.

[25] It is not clear how the Department of Education would operate these merged programs or what their future functional relationship would be to the separate programs as they exist now.

[26] This amount includes funding for five existing programs, including Transition to Teaching, Teacher Quality Partnership, Teachers for a Competitive Tomorrow, Teach for America, and School Leadership.

[27] This amount includes funding for three existing programs, including the High School Graduation Initiative, Advanced Placement, and Javits Gifted and Talented Education.

[28] Telephone and e-mail communications between the author and Patricia Temple, Office of Congressional and Intergovernmental Affairs, U.S. Department of Energy, April 11, 2011.

[29] P.L. 110-69, Title V, Sec. 5003, as amended by P.L. 111-358, Title IX, Sec. 901.

[30] U.S. Department of Energy, Report of the Committee of Visitors of the Office of Workforce Development for Teachers and Scientists (WDTS) in the Department of Energy, May 17-19, 2010, http://science bes/besac/pdf/Wdts_cov_2010_f.pdf.

[31] P.L. 110-69, Title V, Sec. 5009, as amended by P.L. 111-358, Sec. 902.

[32] P.L. 111-358, Title VI, Sec. 603.

[33] P.L. 111-358, Title VI, Sec. 602.

[34] P.L. 111-358, Title VII, Sec. 703.

[35] For more information on the appropriations process, see CRS Report 97-684, *The Congressional Appropriations Process: An Introduction*, by Sandy Streeter.

[36] CRS Report R41721, *Commerce, Justice, Science, and Related Agencies: FY2012 Appropriations*, coordinated by Nathan James, Jennifer D. Williams, and John F. Sargent Jr.

[37] CRS Report R41908, *Energy and Water Development: FY2012 Appropriations*, coordinated by Carl E. Behrens.

[38] U.S. Congress, House Committee on Appropriations, *Energy and Water Development Appropriations Bill, 2012*, report to accompany H.R. 2354, 112th Cong. 1st sess., H.Rept. 112-118 (Washington, DC: GPO, 2011), p. 114. According to the DOE, the fellowship program referenced on page 114 of H.Rept. 112-118 is the Office of Science Graduate Fellowship (SCGF) program. The SCGF program is one of the fellowship programs that the Department asserts is consistent with the PACE fellowship from America COMPETES

2010. The recommended funding level for the SCGF in H.Rept. 112-118 ($5.0 million) is consistent with the FY2010 funding level ($5.0 million) and is $11.1 million (68.9%) below the Administration's FY2012 request for $16.1 million.

[39] Includes $100.0 million in Wireless Innovation (WIN) funds as described in the FY2012 request.

[40] America COMPETES 2010 reauthorizes funding for alignment activities authorized by America COMPETES 2007, Title IV, Subtitle D. Subtitle D activities include grants to states for two activities: (1) to identify the knowledge and skills needed to enter higher education, the armed forces, or the workforce, and to align pre-school-to-college (P-16) education systems with those needs; and (2) to establish or improve statewide P-16 education data systems.

INDEX

C

D

E

F

G

H

I

J

K

L

O

P

Q

R

S

V

W

Y